SKY GAZING

A Guide to the
MOON · SUN · PLANETS · STARS · ECLIPSES · CONSTELLATIONS

Meg Thacher

Storey Publishing

The mission of Storey Publishing is to serve our customers by publishing practical information that encourages personal independence in harmony with the environment.

Edited by Deborah Burns

Art direction and book design by Jessica Armstrong

Text production by Erin Dawson

Indexed by Samantha Miller

Illustrations by © Hannah Bailey

Star wheel information courtesy of the Astronomical Society of the Pacific

Cover photography by Carolyn Eckert (inside back, author); © Dennis di Cicco/Sky & Telescope (back, c.); ESO/S. Brunier (inside, front & back); NASA, ESA, A. Simon (GSFC), and M.H. Wong (University of California, Berkeley) (back, b.r.); NASA Scientific Visualization Studio (back, t.l.)

Interior photography by Adam Block/Mount Lemmon SkyCenter/University of Arizona, 116 t.; Adolf Vollmy/Wikimedia, 115 b.; © Akira Fujii/David Malin Images, 25 l., 89, 97; © Alan Dyer, 91; © AlexaSava/Getty Images, 42; © Ali Candad Photos/Alamy Stock Photo, 30; ALMA (ESO/NAOJ/NRAO)/E. O'Gorman/P. Kervella/Wikimedia, 117 (Betelgeuse); ALMA (ESO/NAOJ/NRAO), S. Andrews et al., N. Lira, 86 r.; © Andrew Hicks/iStock.com, 54; Ann Field/NASA/Wikimedia Commons, 81 (Makemake); © anothersteph/iStock.com, 18 b.; Courtesy of Aubrey Brickhouse at Brickhouse Observatory on the Meyer's Observatory Field, 117 (Antares); © Bill Gozansky/Alamy Stock Photo, 58; © chaphot/stock.adobe.com, 17; © Charly_Morlock/iStock.com, 16 b.r.; © Clayton Fraser, 43 b.; © Dave W. G. Smith, Maldon, UK, 87 t.; © Dennis di Cicco/Sky & Telescope, 55 b.; © Dovapi/iStock.com, 16 b.l.; E. Kolmhofer, H. Raab, Johannes-Kepler-Observatory, Linz, Austria (www.sternwarte.at)/CCommons BY-SA 3.0 Unported/Wikimedia, 85 t.; Erich Karkoschka (University of Arizona) and NASA, 78 b.l.; ESA & NASA/SOHO, 59 b.; ESA/Rosetta/NAVCAM/CC BY-SA IGO 3.0/Wikimedia Commons, 84; ESO/J. Rameau, 86 l.; ESO/Sebastian Deiries, 85 b.; ESO/T. Preibisch, 112 t.; ESO/Y. Beletsky, 64 b.r.; © 2014 by Fred Espenak (MrEclipse.com), 38, 40; © GEOEYE/SCIENCE PHOTO LIBRARY, 68; Gregory H. Revera/CC BY-SA 3.0 Unported/Wikimedia Commons, 71 b.c.r.; Greg Rakozy/Unsplash, 8; Haktarfone at English Wikipedia/CC BY-SA 3.0 Unported/Wikimedia, 117 (Rigel); Howard Perlman/USGS, 71 b.l.; © Imgorthand/iStock.com, 31; © inigofotografia/iStock.com, 43 t.; Jan Sandberg/Wikimedia Commons, 87 b.l.; © Jens Mayer/Shutterstock.com, 23 t.r.; Jeremy Stanley/CC BY 2.0 Generic, 9; Jin Zan/CC BY-SA 3.0 Unported/Wikimedia Commons, 81 (Haumea); JMARS/NASA/Wikimedia Commons, 32 #2, 33 #6; © Jose Antonio Hervás, 63; J.P. Harrington and K.J. Borkowski (University of Maryland), and NASA/ESA, 117 l.; Jstuby at English Wikipedia/CCo/Wikimedia, 34 l.; Kai Pilger/Unsplash, 88; © karinegenest/iStock.com, 16 t.r.; © Kazushi_Inagaki/iStock.com, 28; Khongor Ganbold/Unsplash, 15; Klemen Vrankar/Unsplash, 6, op. 132; © ktasimarr/iStock.com, 18 t.l.; Lawrence Sromovsky, University of Wisconsin-Madison/W.W. Keck Observatory, 78 b.c.l.; © Luis Argerich, 64 l.;

Marcelo Quinan/Unsplash, 23 l.; Mars Vilaubi, 25 r., 49; Mbz1/CC BY-SA 3.0 Unported/Wikimedia Commons, 16 t.l.; © Meg Thacher, 123 l.; Mike Kareh/Unsplash, 44; Morn/CC BY-SA 4.0 International, 118; NASA, 27, 32 #1, 33 t.r. & b.r., 60 b.c.r., 71 t.l.; NASA and The Hubble Heritage Team (STScI/AURA), 76 b.r.; NASA/Aubrey Gemignani, 55 t.; NASA Earth Observatory images by Joshua Stevens, using Suomi NPP VIIRS data from Miguel Román, NASA EPIC Team, 56; NASA, ESA and the Hubble Heritage (STScI/AURA)-ESA/Hubble Collaboration, 112 b.; NASA, ESA, A. Simon (GSFC), and M.H. Wong (University of California, Berkeley), 74 t.; NASA/ESA/Hubble Heritage Team (STScI/AURA), 52 t.; NASA/ESA/M. Robberto (STScI/ESA) and the Hubble Space Telescope Orion Treasury Project Team, 52 b.; NASA/GSFC/Arizona State University, 32 #3; NASA/GSFC, 10–11, 51 t.l.; NASA/GSFC/CC BY 2.0/Wikimedia Commons, 51 b.; NASA/GSFC/SDO/Genna Duberstein, 64 t.r.; NASA/GSFC/University of Arizona, 67; NASA/GSFC/LaRC/JPL, MISR Team, 71 b.c.l.; NASA/JHUAPL/SwRI, 80 b.c.; NASA/Johns Hopkins University Applied Physics Laboratory/Carnegie Institution of Washington, 60 b.r., 69; NASA/Johns Hopkins University Applied Physics Laboratory/Southwest Research Institute, 80 all ex. b.c.; NASA/JPL, 60 t.l. & m.r., 70 t., b.l., b.c.l & b.c.r., 72 b.l., 78 b.c.r., 79 t., b.l. & b.r.; NASA/JPL-Caltech, 60 t.c., t.r. & b.c., 70 t., 72 t. & b.c.l., 75 t.r., 78 t., 79 b.c.l.; NASA/JPL-Caltech/University of Arizona, 72 b.c.r & b.r.; NASA/JPL-Caltech/MSSS, 73; NASA/JPL-Caltech/SETI Institute, 75 b.r.; NASA/JPL-Caltech/Space Science Institute, 76 b.l., b.c.l. & b.c.r., 77 b.r.; NASA/JPL-Caltech/SwRI/ASI/INAF/JIRAM, 74 b.r.; NASA/JPL-Caltech/SwRI/MSSS/Kevin M. Gill, 60 b.l.; NASA/JPL-Caltech/SwRI/MSSS/Kevin M. Gill/CC BY 3.0 Unported, 74 b.l.; NASA/JPL-Caltech/UCLA/MPS/DLR/IDA, 81 (Ceres); NASA/JPL/Cornell University, 74 b.c.r.; NASA/JPL/DLR, 75 t.l.; NASA/JPL/ESA/University of Arizona, 77 m.; NASA/JPL/Space Science Institute, 76 t., 77 b.l.; NASA/JPL/University of Arizona, 74 b.c.l., 75 b.l.; NASA/JPL/University of Arizona/University of Idaho, 77 t.; NASA/JPL/USGS, 78 b.r., 79 b.c.r.; NASA's Scientific Visualization Studio, 26, 32–33, 36, 37, 41, 123 r.; Courtesy of NASA/SDO and the AIA, EVE, and HMI science teams, 51 t.r.; NASA Visualization Technology Applications and Development (VTAD), 81 (Eris); NASA/Wikimedia, 34 r.; NASA (image by Lunar Reconnaissance Orbiter)/Wikimedia Commons, 32 #4, 33 #5; Natarajan.Ganesan/CC BY-SA 4.0 International/Wikimedia, 55 m.; Olaf Tausch/CC BY3.0 Unported/Wikimedia, 7; © Paul Souders/Getty Images, 47; © Paul Wilson/iStock.com, 18 t.r.; Philippe Donn/Pexels, 21; Philipp Salzgeber/CC BY-SA 2.0 Austria/Wikimedia Commons, 87 b.r.; © Riccardo Beretta/iStock.com, 29; © Science History Images/Alamy Stock Photo, 65; Courtesy of Sebastian Saarloos, 23 b.r.; © SellOnlineMarketing/iStock.com, 48; SOHO (ESA & NASA), 59 t.; Stephen Rahn/Wikimedia Commons, 117 (Vega); Strebe/CC BY-SA 3.0 Unported/Wikimedia, 57; Image © Ted Stryk. Data Courtesy the Russian Academy of Sciences, 70 b.r.; Visualization by Earth Observatory, 71 b.r.; © wisanuboonrawd/iStock.com, 115 t.; © www.capella-observatory.com, 116 b.; © Yuri Beletsky, op. title page, 61

Storey books are available at special discounts when purchased in bulk for premiums and sales promotions as well as for fund-raising or educational use. Special editions or book excerpts can also be created to specification. For details, please call 800-827-8673, or send an email to sales@storey.com.

Storey Publishing
210 MASS MoCA Way
North Adams, MA 01247
storey.com

Printed in China by R. R. Donnelley
10 9 8 7 6 5 4 3 2 1

Library of Congress Cataloging-in-Publication Data

Names: Thacher, Meg, author.

Title: Sky gazing : a guide to the moon, sun, planets, stars, eclipses, constellations / Meg Thacher.

Description: North Adams, MA : Storey Publishing, [2020] | Includes bibliographical references and index. | Audience: Ages 9–14 | Audience: Grades 4–6 | Summary: "With this highly visual guide to observing the sky with the naked eye, kids aged 9–14 will delve into the science behind what they see"— Provided by publisher.

Identifiers: LCCN 2020016777 (print) | LCCN 2020016778 (ebook) | ISBN 9781635860962 (hardcover) | ISBN 9781635860979 (ebook)

Subjects: LCSH: Astronomy—Observers' manuals—Juvenile literature.

Classification: LCC QB64 .T45 2020 (print) | LCC QB64 (ebook) | DDC 520—dc23

LC record available at https://lccn.loc.gov/2020016777

LC ebook record available at https://lccn.loc.gov/2020016778

What's in This Book?

With most books, you start at the beginning and read through to the end. This book was made to skip around in. The chapters are organized from closest (the Moon) to farther (the Sun and planets) to farthest (the stars).

Hi! Call me Star Dude. I'll show up now and then to guide you through this book and share some things I know.

1
Step into the SKY

Your Astronomy Notebook will help you keep track of your discoveries.

2
the MOON

Each chapter ends with a section called A Closer Look on how to observe the sky with binoculars.

Some activities are easy ★ and some are harder ★★★. Follow the stars!

3

the SUN

4

PLANETS

You'll also find instructions for viewing Special Events like eclipses and meteor showers.

If you want to start learning constellations right away, you'll find sky maps here.

5

STARS and CONSTELLATIONS

APPENDIX:
Find Out More

The glossary has definitions of words you may not know — and words that astronomers use in a different way from most people. (In the book those words are highlighted, like **this**, when first used.)

Imagine that the solar system could fit in this book. If the Sun could fit on this page, what page would each planet be on, approximately? See if you can find them all! (Answers on page 131.)

Sun

Step into the

SKY

Our universe is filled with stars, planets, and all sorts of amazing stuff — and you can see them no matter where you live. You don't need fancy tools: just look up.

THE SKY BELONGS TO EVERYONE

Long ago, kids knew all about the night sky. They could find north and tell time by the Sun. They knew which constellations came with which season.

Of course, this was easier before the invention of streetlights. The sky they saw was speckled with thousands of stars. These days, we can see only a few hundred from our cities and suburbs. Many people live their whole life without seeing the Milky Way.

But no matter where you live or how many stars you can see at night, you can observe the sky. You can do everything those kids from long ago could do — and more! We've learned a lot about the universe since our ancestors started sky gazing.

This book is about astronomy, the study of stars, planets, and space. Astronomy is interesting for its own sake, but it's also an important part of human history. Ever since there were people, we've been looking at the stars: tracking and recording their motion, making pictures and stories out of them, and wondering *why* they are there.

The sky inspired us to invent math and physics so we could explain what caused nature's patterns, starting with how objects move across the sky. It got us thinking about more than just what to eat and where to live; it showed us our place in the universe.

You can observe the night sky anytime, anywhere — for free! Start a habit of looking up at stars whenever you step outside at night.

Part of the Egyptian zodiac from the Temple of Hathor in Dendara, Egypt, built around 50 BCE. It was sculpted and painted onto the temple's ceiling. The body of Nut, the goddess of the sky, lies along the bottom.

Looks like Venus has been waiting for us!

SKY-GAZING SUPPLIES

If you want to watch a meteor shower or have a sky-gazing party, pack some extra stuff to keep yourself comfortable outside at night.

You might want to bring:
- ★ water and a snack
- ★ a sky map or Star Wheel (see chapter 5)
- ★ a blanket
- ★ a red flashlight (see page 19)
- ★ a regular flashlight
- ★ bug repellent
- ★ a pencil
- ★ your Astronomy Notebook (see page 8)

Or you can just step out onto your fire escape or porch and look up!

For more tips on throwing a star party, see page 119.

WHAT'S UP THERE?

No matter how dark or light your sky is, you can always observe the Sun and Moon! And even from a lit-up place like a city or large suburb, you can see the brightest planets, stars, and **meteors** (flashes of light caused by bits of rock from outer space entering the Earth's atmosphere), and even the International Space Station.

From a darkish place outside the suburbs, you can see most of the **constellations** (groups of stars that look like a picture). You can also see meteors and human-made satellites. If it's dark enough, you may see a faint trace of the **Milky Way**, the galaxy we live in. (A **galaxy** is a huge star system containing gas, dust, and hundreds of billions of stars.)

If you have the chance to visit a place that is very dark at night, like a national park, you will see the Milky Way clearly, with its many stars and dust lanes. Star clusters, **nebulae** (clouds of gas and dust), and even galaxies may be visible as well.

 Space Talk The word *nebula* comes from the Latin word for cloud. One is a nebula; two or more are nebulae.

START A SKY JOURNAL

When you sky gaze, it's fun to draw what you see in the sky and take notes.

An Astronomy Notebook will keep all of your observations together. You can buy a new notebook or use an old one that still has some blank pages left. Rip out your old schoolwork, and decorate the front.

Use your Astronomy Notebook to record the weather, draw pictures of the sky, and keep track of your research.

When you record an observation you should include the following information:

→ **Date**

→ **Time**

→ **Weather**

→ **How clear is the sky?**

→ **Can you see the Moon? Where is it and what shape?**

→ **What else can you see?**

→ **What's different from the last time you looked?**

Venus

WHERE TO SKY GAZE

Good places to look at the stars include a backyard, a balcony, a safe rooftop that you're allowed to go out on, a playing field, or a park. Try to find a place with a clear **horizon** — that is, where nothing blocks your view. Maybe you can see the southern and western sky from where you live, and the north and east from a friend's house. Observing from a hill or rooftop can get you a better view.

In modern times, we have many lights inside and outside our homes. They help us see at night, but they can cause **light pollution** (artificial light that makes the sky bright).

The constellations of Sagittarius and Scorpius, as seen from a city (population 400,000) and a small town (population 217).

A handy way to block point-source light pollution

Too Much Light

If the sky is brighter than a particular star, we can't see that star. So in places with a lot of light pollution, we can see only really bright stars and planets.

DIFFUSE light pollution lights up the whole sky. You can't get away from it unless you go somewhere with less light.

POINT-SOURCE light pollution comes from one place. You can move to get away from it. If a nearby streetlight is making it hard to see the stars, for example, you can move so that a building blocks it from your view, or you can hang a blanket from a clothesline to block it out. Even just blocking the light with your hand will improve what you see!

Best Timing

Once you've found a good place to observe the night sky, keep the following things in mind.

★ **WEATHER.** You can observe the Moon and planets through a partly cloudy sky. To see constellations well, you need an almost clear sky.

Be sure to dress appropriately. Weather is hard to predict, so check the sky and the temperature before you head out to make your observations!

★ **TIMING.** You'll have to wait until twilight — the time right after sunset or before sunrise when the sky is not bright but still softly glows. Then it will be dark enough to observe most stars.

★ **MOON.** Check the Moon's phase (that is, how much of the Moon is visible from Earth; see page 27) and its rise and set times. If it's a very bright Full Moon, it will outshine fainter objects in the sky and you won't be able to see them. (Observe the Moon instead!)

★ **PLANETS.** Often you may see a planet that is brighter than most stars. Look online or in an astronomy magazine to learn what planets will appear in the sky and when. Over several months you can watch planets change positions as if they're in a big dance. (See Chapter 4.)

★ **SPECIAL EVENTS.** Eclipses, meteor showers, comets, the northern lights — there's always something interesting going on in the sky! Special events are found in every chapter of this book, and many are listed in the Appendix, "Find Out More."

Earth

DARKNESS AND LIGHT

Lighting is important at night; we need to see where we're going and feel safe. Unfortunately, the light we use doesn't shine just on streets or buildings or people. Some of it shines up or out, instead of down. It comes from streetlights, buildings, playing fields, and security lights.

The International Dark-Sky Association (IDA), listed in the resources, helps raise awareness about light pollution and how to improve lighting. Reducing artificial light at night can make us and our environment happier and healthier. And we'll all get to see more in the clear, dark sky.

Measuring Darkness

Astronomer John Bortle has classified light pollution on a scale from 1 to 9.

Bortle class 1 skies are found in the darkest places on Earth, away from all light. On nights with no Moon, the Milky Way and bright nebulae and star clusters are visible with the naked eye.

Bortle class 9 skies are found in big cities. You can see the Moon and planets, but only a few bright stars — not enough to recognize their constellations.

BORTLE SCALE	MAGNITUDE WE CAN SEE	TYPE OF PLACE
1	7.6	Dark sky preserve*
2	7.1	Wilderness or national park, with some light on the horizon*
3	6.6	State or national park near a small town*
4	6.1	Small town
5	5.6	Suburb
6	5.5	Large suburb
7	5	Edge of a city
8	4.5	City
9	4	Center of large city

*Most people's eyes can't see any difference between Bortle 1, 2, and 3 skies. You will notice the difference when you look around, though. At very dark sites with no Moon, you can't see the person next to you unless they have a flashlight!

Space Talk A star's magnitude tells us how bright it is. It may seem backward: faint stars have large magnitudes; bright stars have small magnitudes. See page 89 for more information.

NASA's "Black Marble," an image of the Earth at night, seen from space in 2012. The dark areas (like oceans and deserts) are places with little or no light pollution.

Mars

FINDING YOUR WAY AROUND THE SKY

When you see something cool way up above you, how can you help someone else find it? Scientists use special words to describe places in the sky.

ZENITH
An imaginary point straight overhead

MERIDIAN
An imaginary line across the sky, from the north point on the horizon through the zenith to the south point on the horizon

ALTITUDE
An object's angle of elevation (height) above the horizon

AZIMUTH
The object's angle around the horizon from one of the cardinal directions (north, south, east, and west)

Moon

EAST

NORTH

SOUTH

HORIZON
An imaginary circle around you, at the height of your eyes

WEST

Space Talk **meridian:** divides the sky into an eastern (rising) half and a western (setting) half. Objects in the sky reach their highest point when they cross, or *transit*, the meridian.

NADIR
An imaginary point opposite the zenith — right beneath your feet

How It Looks to Us

From Earth, the sky resembles the inside of a giant upside-down bowl. The Sun, Moon, planets, and stars all move from east to west across the sky.

Of course, this is only what the sky *looks like*...

But What's Really Happening?

In reality, we live on a ball-shaped planet that spins from west to east around an imaginary center line called an **axis**. The Earth makes one complete spin every 24 hours. As we spin from west to east, the Sun, Moon, and stars seem to move from east to west.

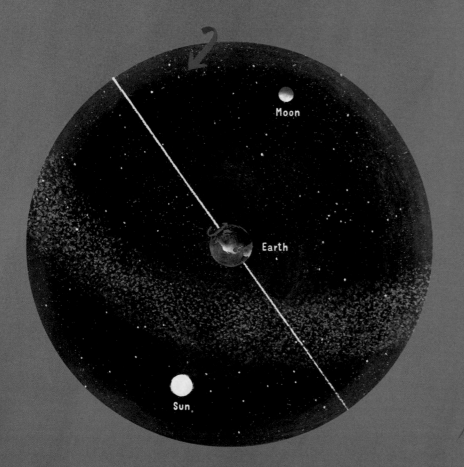

Moon

Earth

Sun

COSMIC PROTRACTOR ✦✦

Astronomers measure distances by picturing angles on the dome of the sky.

For schoolwork you might use a protractor to measure angles, but it won't work for measuring angles on the sky. For that, you need a cosmic protractor.

You have your own measuring tools that you always carry with you: your hands and fingers! Hold your hand up to the sky with your arm straight to make these measurements.

1 degree

Twice the size of the full Moon; size of the Pleiades (Seven Sisters)

2 degrees

A little smaller than Orion's belt

10 degrees

Handle of the Big Dipper

15 degrees

Cassiopeia

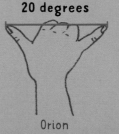

20 degrees

Orion

BE A WEATHER WATCHER ★

Keep a record of the weather so you know the best time for sky gazing.

Describe the weather in your notebook whenever you can. Try to record the weather once during the day and once at night. Here are some things you can record. Don't forget the date!

→ **What is the temperature?**

→ **Is the sky clear or cloudy?**

→ **Is it very windy?**

→ **Is there fog?**

→ **How much of the sky is covered by clouds?**

→ **What kinds of clouds do you see? (You can draw them.)**

→ **Is it raining or snowing?**

→ **If it is, is the rain or snow light, heavy, or somewhere in between?**

After you've been keeping your weather journal for a few months, look for patterns. Do you notice that some months are cloudier than others? If the sky is cloudy during the day, is it also cloudy at night? Does rain always come with certain types of clouds?

Where Clouds Live

The names for clouds are all combinations of the following five Latin words.

CIRRUS: hair

CUMULUS: pile or heap

STRATUS: layer

NIMBUS: rain cloud or mist

ALTO: high

HIGH
20,000 feet (6,000 m) or more above the ground

MIDDLE
6,500–20,000 feet (2,000–6,000 m) above the ground

LOW
Less than 6,500 feet (2,000 m) above the ground

Cirrocumulus

Cirrostratus

Cirrus

Cumulonimbus
(can reach as high as 50,000 feet — that's 9 miles, or 15 km!)

Altocumulus

Altostratus

Nimbostratus

Cumulus

Stratocumulus

Stratus

LIGHT SHOWS

The light from the Sun looks white, but it's made of all colors. We can see this when there's a rainbow. Rainbows happen when it's sunny and rainy at the same time. The sunlight bounces around inside the raindrops, and different colors of light are bent or refracted at different angles in the water. When the light comes out of the drop the red, orange, yellow, green, blue, and purple light end up in different places.

The raindrops form a cone of colored light, so rainbows are really full circles! We only see the part of the rainbow that is above the horizon.

The best time to see a rainbow is when the Sun is low on the horizon in the morning or evening. If it's raining and the Sun is out, turn your back to the Sun and look for the rainbow.

To make your own rainbow, use a garden hose or fountain as your rainstorm. Look at the spray of water with your back to the Sun. You should see a rainbow to the left and right of where you're facing, floating in the air. Home-made rainbows show up best on a dark background.

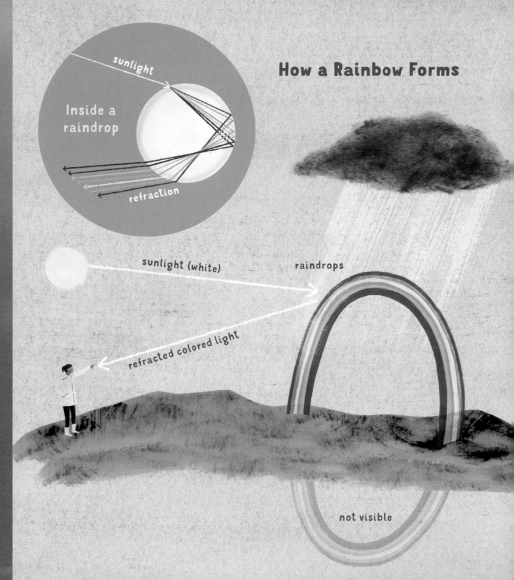

How a Rainbow Forms

Inside a raindrop

sunlight

refraction

sunlight (white)

raindrops

refracted colored light

not visible

Double rainbows happen when the sunlight is pretty bright. The light bounces twice inside some raindrops and makes another rainbow outside the main one. The second bow will be fainter, with the colors in reverse order.

SPECIAL EFFECTS

The Sun is like an artist! Its light can play with clouds to make many beautiful effects. Be careful not to look directly at the Sun.

A SUNDOG is a rainbow image of the Sun to its right or left. The sunlight bounces off ice crystals in cirrus or cirrostratus clouds in the same way a rainbow is formed. The clouds act like a screen where the Sun images are projected. If you see a sundog, look for a halo (below)!

A SUN PILLAR occurs when light from the Sun reflects off ice crystals directly above it in a pillar shape.

CREPUSCULAR (TWILIGHT) RAYS are caused by clouds blocking the Sun's light in some places and letting it through in others.

A HALO can form around the Sun from cirrostratus clouds. Try blocking the Sun with your hand. The halo will be outside your fingers. Moon haloes are much easier to see than Sun haloes.

Why Are Sunsets Red?
Because the Sky Is Blue!

When sunlight passes through Earth's atmosphere, its blue light particles collide with air molecules and scatter off in all directions. On a sunny day, the sky looks blue.

At sunrise and sunset, the sunlight has to pass "sideways" through more atmosphere than in daytime, so less blue light makes it to our eyes. Without the blue light, the sun looks red or orange.

At sunset and sunrise, with more atmosphere between the Sun and us, we see less blue light and more red.

At midday, with less atmosphere between the Sun and us, we see white sunlight and blue sky.

On the Moon, where there is no atmosphere, the sky is never blue, red, or white, but always black.

WATCH DAY TURN INTO NIGHT ★

For astronomers, sunset is the most exciting time of day. It's the beginning of the night, when the stars and planets become visible.

For this activity, you will need a clear view of the western horizon. Check a weather website to find out when the Sun will set. Half an hour before sunset, head outside and find a comfortable seat facing west.

Watch the clouds and the sky colors change as the Sun slowly sinks below the horizon. Notice how it still lights up the sky and clouds for a while, even after it has set.

Take pictures, or draw what you see. Listen to day sounds around you changing to night sounds.

Earth's shadow rising

Green flash

If you're with friends, see who can spot the first star. How long after sunset do you have to wait?

The Anti-Sunset

On very clear nights, watch the *eastern* (not western!) horizon after sunset. You may see the "twilight wedge," or anti-sunset, a rising wall of darkness as the Earth's shadow rises into the eastern sky.

The Green Flash

If you have a wide-open view of the horizon (over the sea or desert, for example) and the weather is clear, you may be able to see the "green flash." This is a green spot of light that can appear on the upper edge of the Sun just before it sets. It's pretty rare and lasts for only a second or two. (Never look directly at the Sun, even as it sets.)

WEST ←

This panoramic photo shows the whole horizon. The sunset is on the left, and the dark blue twilight wedge is on the right.

→ EAST

NIGHT VISION

Your eyes can adjust to the dark in 5 or 10 minutes, but it will take up to 30 minutes to get your best night vision. If your eyes have adapted to the dark but then are suddenly exposed to a bright light, they'll have to begin adjusting all over again, though it probably won't take as long.

Your eyes react differently to different colors of light. Red light does not affect your night vision, while bright blue light — like the light from a cell phone or computer screen — can damage it. You can download apps that make your phone or computer screen less blue and more red to help preserve your night vision.

Keep your phone in your pocket when you're not using it. Better yet, put your screens away until after you are done sky gazing. This helps you sleep better, too!

HOW WE SEE

Cornea

Iris

Retina

Pupil
Light comes into the eye through the pupil.

Lens
The lens bends the light to form an upside-down image on the retina. Your brain flips the image so you can see right-side-up.

Optic Nerve
The optic nerve transmits information from the retina to your brain, which tells you what you're seeing.

MAKE A RED FLASHLIGHT

A red flashlight helps you see your surroundings but protects your night vision.

YOU WILL NEED:

- Red balloon or red plastic wrap
- Flashlight
- Marker
- Scissors
- Rubber band, hair elastic, or duct tape

1. Place the balloon flat on a table.

2. Set the flashlight down on the balloon with its face (the round end that shines) flat on the balloon.

3. Trace around the face of the flashlight with the marker.

4. Use the scissors to cut out a circle from the balloon that is about 1 inch (2.5 cm) bigger all around than the circle you just traced.

5. Stretch the red circle over the front of the flashlight, and secure it with an elastic band or tape.

6. Try out the flashlight at night, after your eyes have adapted to the dark. If the flashlight seems too bright, add another red layer.

You can also tape over the front of the flashlight with red masking tape or duct tape. Or, if your flashlight has a standard bulb (not a halogen or LED bulb), you can take the flashlight apart and paint its light bulb with red nail polish. Orange also works well.

Journey into Our Home Galaxy
THE MILKY WAY

All the stars you can see in the sky belong to our Milky Way galaxy. The Milky Way contains about 250 billion stars, as well as gas and dust.

There are about 200 billion galaxies in our universe — 2 trillion, if you count all the baby galaxies that haven't yet smashed together to make full-grown galaxies.

Can We See Other Galaxies?

We can see three galaxies from Earth without a telescope, not counting our own Milky Way. They are the Andromeda Galaxy in the Andromeda constellation, the Large Magellanic Cloud (LMC) in Dorado, and the Small Magellanic Cloud (SMC) in Tucana. The LMC and SMC are visible only from the Southern Hemisphere.

Four other galaxies are visible only on clear nights with little or no light pollution: the Triangulum Galaxy in Triangulum, Centaurus A in Centaurus, Bode's Galaxy in Ursa Major, and the Sculptor Galaxy in Sculptor.

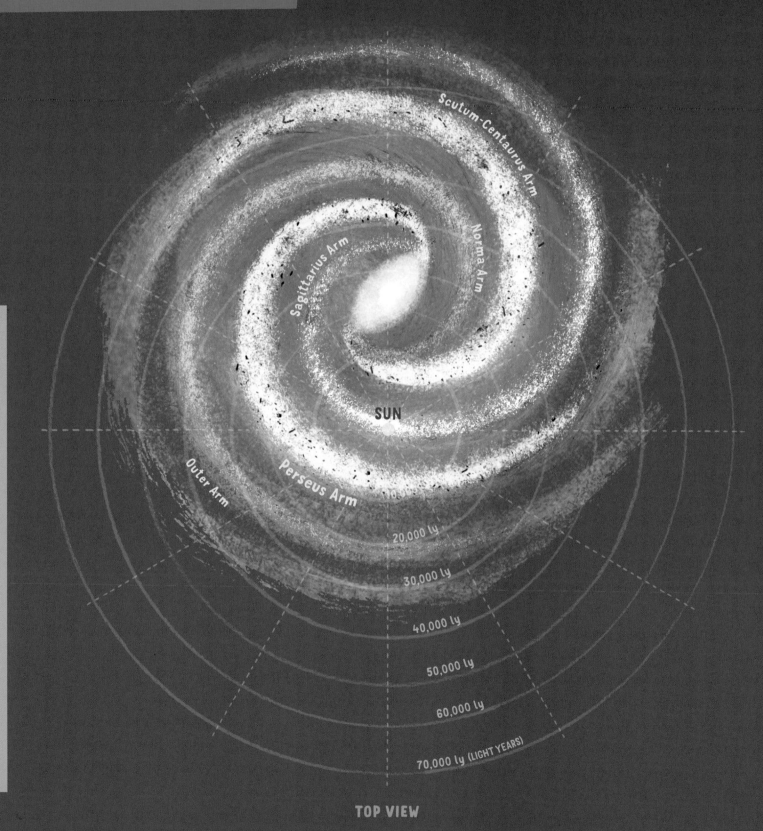

Scutum-Centaurus Arm

Norma Arm

Sagittarius Arm

SUN

Outer Arm

Perseus Arm

20,000 ly

30,000 ly

40,000 ly

50,000 ly

60,000 ly

70,000 ly (LIGHT YEARS)

TOP VIEW

How It Looks to Us

The Milky Way is a cloudy stripe across the sky, like a silver river. You can see it on a clear, dark night. It's brightest in the constellation of Sagittarius.

But What's Really Happening?

The Milky Way is a very thin disk, about 100,000 light-years in diameter and 1,000 light-years thick. It has spiral arms and a dense bulge at its center.

When we look along the disk of the galaxy, we see lots of stars — so many that they look like a milky haze. When we look out of the disk, we see fewer stars.

Space Talk A *light-year* is how far light travels in a year. It's about 5.88 trillion miles, or 9.46 trillion km.

VIEW FROM EARTH

Some stars

Lots of stars

Dark patches within the Milky Way are dust blocking the starlight.

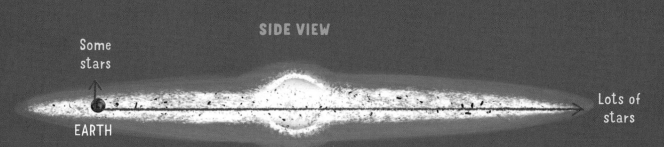

SIDE VIEW

Some stars

EARTH

Lots of stars

The Milky Way galaxy is disk-shaped. When we look up out of the disk, we see some stars. When we look along the disk, we see lots of stars.

THE AURORA

The **aurora** is the beautiful northern and southern lights that people who live near the North and South Poles can see. The aurora can look like glowing clouds, curtains blowing in the wind, dancing serpents, or streamers of light.

Polar lights are caused by the **solar wind** — a steady stream of charged particles flowing from the Sun. When they reach Earth, some of these particles are drawn in to flow along the Earth's magnetic field. They collide with **atoms** in the atmosphere, making them glow.

Let's Dance!

Electrons hit atoms in the air

Atoms get excited

Atoms give off light as they calm down

The aurora is not just pretty lights in the sky. It's a sign that the Earth's magnetic field is protecting us from the solar wind.

Solar wind

Solar wind particles

Solar wind particles

Solar wind

MAGNETIC FIELD

Particles from the solar wind bend around our planet or spiral into the atmosphere to produce the aurora.

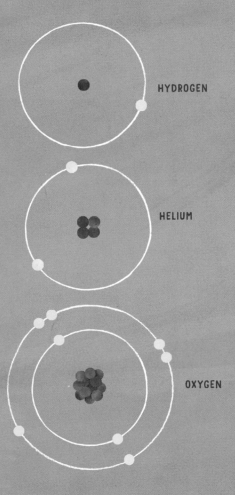

Atoms

The ordinary matter around us is made of atoms, tiny particles that are too small to see. Inside each atom is an even tinier **nucleus** surrounded by ⬤ **electrons**, and inside the nucleus are ⬤ **protons** and ⬤ **neutrons**. Protons have a positive electrical charge, electrons have a negative charge, and neutrons are neutral.

HYDROGEN

HELIUM

OXYGEN

The number of protons in an atom's nucleus tells you what kind of atom it is. For example, hydrogen has one proton, helium has two protons, and oxygen has eight protons (two are hiding behind the others!).

How to Observe the Aurora

The best time to see the aurora is on a winter night with no moon. Not because the aurora happens only in winter, but because the nights are longer then, so you have a better chance of seeing it.

You can find out when the aurora is likely to happen from space weather websites. At some of these sites you can even sign up for an alert system. If you receive an alert, go outside, let your eyes adjust to the darkness, and look toward the northern horizon in the Northern Hemisphere, southern horizon in the Southern Hemisphere. If you see a glow that is not a nearby city or town, it's probably the aurora.

Dress warmly and watch for a while. The aurora changes shape and color as different levels of the atmosphere are energized.

The closer you are to one of the Earth's magnetic poles, the more likely you are to see the aurora.

The aurora comes in many colors. Yellow and green occur when electrons collide with oxygen atoms. Red, violet, and – very rarely – blue occur when electrons collide with nitrogen atoms.

Binoculars

This book is mostly about observing the sky with your naked eye. But if you have a pair of binoculars, they can bring the sky to life. See page 123 for directions on binocular care.

Focusing

Most larger binoculars have a focus knob between the two barrels. In addition, many have a ring around one of the eyepieces that lets you focus it separately, since most people have slightly different vision in each eye.

1. To focus, aim the binoculars at the stars.

2. Cover the **objective** lens that's attached to the adjustable eyepiece with your hand, being careful not to touch the lens.

3. Turn the focus knob until your view gets clearer. The stars should be tiny pinpoints.

4. Uncover the first objective lens, cover the other, and now focus this side using only the eyepiece focuser.

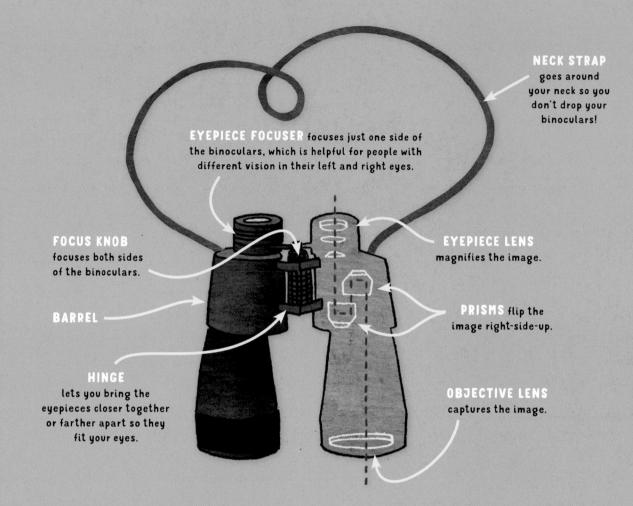

NECK STRAP goes around your neck so you don't drop your binoculars!

EYEPIECE FOCUSER focuses just one side of the binoculars, which is helpful for people with different vision in their left and right eyes.

FOCUS KNOB focuses both sides of the binoculars.

BARREL

HINGE lets you bring the eyepieces closer together or farther apart so they fit your eyes.

EYEPIECE LENS magnifies the image.

PRISMS flip the image right-side-up.

OBJECTIVE LENS captures the image.

Smaller binoculars often have two eyepiece focusers and no main focuser. Focus the eyepieces one at a time.

If you wear glasses, you can look through the binoculars without your glasses on, and adjust the focus. If you want to keep your glasses on, fold back the eye guards.

If your binoculars fog up, do *not* try to wipe them off! This can scratch the lenses. Instead, use a lens blower or a blow dryer to clear the lenses.

USING AN EYE GUARD

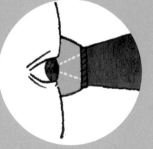

An eye guard blocks stray light and places your eyes the right distance from the eyepiece.

If you wear glasses, you can fold back the rubber eye guard.

Pointing and Finding

When observing, give your eyes time to adapt to the dark. Start with bright objects, and work down to fainter objects as your eyes become more sensitive.

Finding what you're looking for with binoculars can be hard because they limit the amount of sky you can see. Here are tricks to improve your skills.

★ **EYES FIRST.** One good method is to stare at an object with your naked eye, then slooooowly raise your binoculars to your eyes. That will usually get you pretty close.

★ **SPIRAL SEARCH** If you can't see the object you're looking for with binoculars, try moving the binoculars in a spiral — in larger and larger circles — from where they're pointed.

★ **VERTICAL CLIMB.** Another trick is to point your binoculars at an object on the horizon directly below what you want to look at, and move them slowly up.

Spiral Search

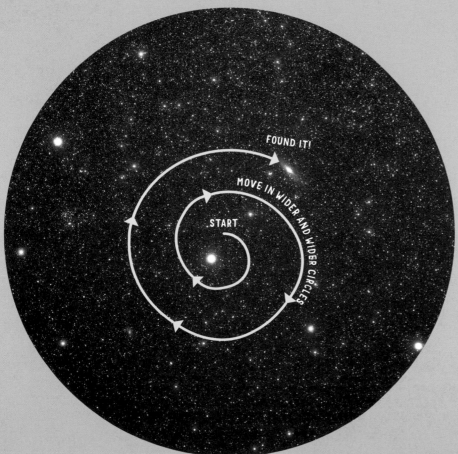

FOUND IT!

MOVE IN WIDER AND WIDER CIRCLES

START

To keep your binoculars steady, lean your elbows on a fence or wall. Use a tripod for binoculars with a magnification greater than 10. If your object is high up in the sky, lie down on the ground or lean back in a beach chair to avoid getting a pain in the neck!

Jupiter

the MOON

The Moon is the second-brightest object in the sky. (The brightest is the Sun, of course!)

SHAPE-SHIFTER

Probably the most noticeable thing about the Moon is that it seems to change shape from one night to the next. The different shapes of the Moon are called phases.

It takes about a month for the Moon to go through all of its phases. In fact, the word *month* comes from the Old English word for moon.

How Do Phases Happen?

The Moon doesn't really change shape — it's always a sphere. What we see is the part of it that is lit up by the Sun. As the Moon orbits Earth, different parts of it are lit up, and so it looks different to us from Earth.

If there were people on the Moon, they would see Earth go through phases, too. Earth's phases are the opposite of the Moon's, and from the Moon, the Earth looks four times as large as the Moon looks from Earth.

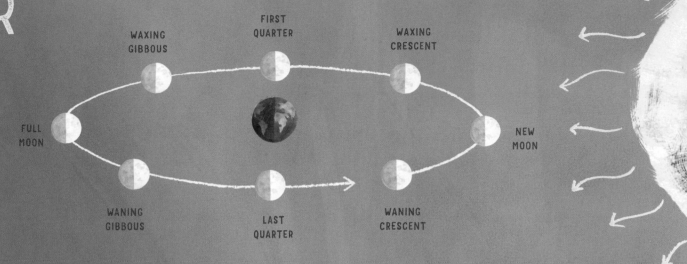

	FIRST QUARTER	
WAXING GIBBOUS		WAXING CRESCENT
FULL MOON		NEW MOON
WANING GIBBOUS	LAST QUARTER	WANING CRESCENT

Phases of the Moon as seen from Earth in the Northern Hemisphere

| NEW MOON | WAXING CRESCENT | FIRST QUARTER | WAXING GIBBOUS | FULL MOON | WANING GIBBOUS | LAST QUARTER | WANING CRESCENT | NEW MOON |

People in the Southern Hemisphere see the Moon upside down compared to how people see it in the Northern Hemisphere.

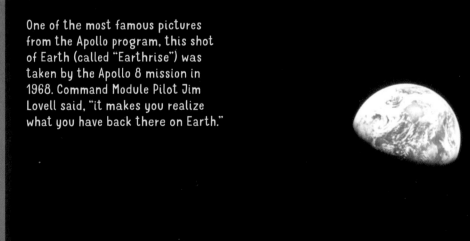

One of the most famous pictures from the Apollo program, this shot of Earth (called "Earthrise") was taken by the Apollo 8 mission in 1968. Command Module Pilot Jim Lovell said, "it makes you realize what you have back there on Earth."

ASTRONOMY NOTEBOOK
MAKE A MOON DIARY ★

The Moon looks slightly different every day. You can track these changes in a Moon Diary.

You can start your Moon Diary anytime, except when the Moon is new and can't be seen. Look for the Moon during the night *and* during the day. When you see it, record in your notebook:

MY MOON DIARY
Dec 17, 7:43pm
Altitude: setting
Direction:
west-Southwest

★ The time and date

★ Where the Moon is in the sky (how high and in what direction)

★ A sketch of the Moon's shape

Look up every few hours to find the Moon. If you can, write down the time that it sets. Watch the Moon every day when the sky is clear.

→ **What are some of the things you notice about the Moon?**

→ **How does its shape change?**

→ **Is it always up at the same time of day or night?**

When the Moon is a crescent, the gibbous Earth reflects the Sun's light onto the craters on its dark side. This is called earthshine.

OUR ONE AND ONLY MOON

AVERAGE DISTANCE FROM EARTH: 238,900 miles (384,472 km)

DIAMETER (compared to Earth): 0.27 (about the size of the United States)

MADE OF: rock

ATMOSPHERE: none

GRAVITY: 1/6 of Earth gravity

TEMPERATURE: -414°F (-248°C) on the dark side; 253°F (123°C) on the bright side

DAYS TO ORBIT EARTH: 27.32

DAYS IN PHASE CYCLE: 29.53

You may notice that the phase cycle is a little longer than the time it takes the Moon to orbit the Earth. That's because while the Moon is orbiting the Earth, the Earth is orbiting the Sun. To get back to the new phase, the Moon has to travel another 2.2 days in its orbit.

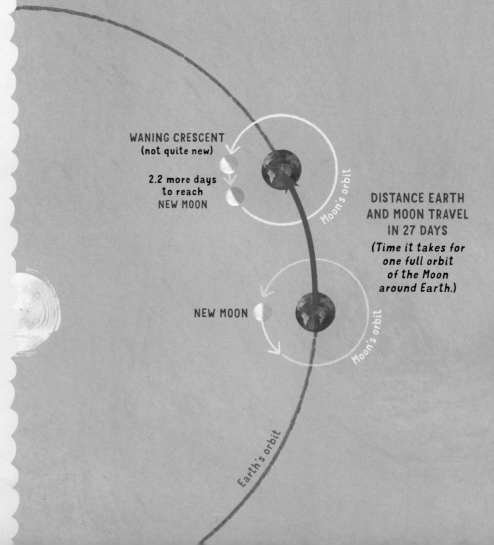

WANING CRESCENT (not quite new)

2.2 more days to reach NEW MOON

Moon's orbit

DISTANCE EARTH AND MOON TRAVEL IN 27 DAYS
(Time it takes for one full orbit of the Moon around Earth.)

NEW MOON

Moon's orbit

Earth's orbit

MOONRISE, MOONSET

Although the Moon always travels from east to west, it takes a different path across the sky every day. It rises about 50 minutes later every day until its cycle is complete, in about a month.

The full Moon is opposite the Sun. It rises at sunset and sets at sunrise. The new Moon rises with the Sun at sunrise and sets with it at sunset.

Looking due west at sunset five days in a row, the Moon is a little higher and a little fuller each day. (In this simulated image, the size of the Moon is exaggerated.)

PHASE	RISES*	TRANSITS	SETS*
NEW	6:00 a.m.	Noon	6:00 p.m.
WAXING CRESCENT	9:00 a.m.	3:00 p.m.	9:00 a.m.
FIRST QUARTER	Noon	6:00 p.m.	Midnight
WAXING GIBBOUS	3:00 p.m.	9:00 p.m.	3:00 a.m.
FULL	6:00 p.m.	Midnight	6:00 a.m.
WANING GIBBOUS	9:00 p.m.	3:00 a.m.	9:00 a.m.
THIRD QUARTER	Midnight	6:00 a.m.	Noon
WANING CRESCENT	3:00 a.m.	9:00 a.m.	3:00 p.m.

*Rise and set times are approximate (transit times are exact).

Space Talk **Transit:** Cross; as when a celestial body crosses the meridian, or crosses another celestial body, such as the face of the Sun.

DAY 5

DAY 4

DAY 3

DAY 2

DAY 1

MOON ILLUSION

When the full Moon is on the horizon, it looks bigger than when it's high in the sky. But it's really the same size no matter where you see it. You can prove this to yourself by measuring the Moon with your cosmic protractor (see page 13) when it's on the horizon and when it's high in the sky.

So why does the Moon look bigger on the horizon? We don't know! It's an optical illusion — a trick that our eyes play on our brain. There are two possible explanations.

1. The sky doesn't really look like a hemisphere. It looks slightly flattened, with the zenith closer to us than the horizon. This makes the Moon seem farther away — and bigger — when it's on the horizon.

2. There are landmarks on the horizon. We see the Moon next to large objects like buildings and trees and think the Moon is large, too. When it's at the zenith, there are no landmarks, so it looks smaller.

What's a Supermoon?

The Moon's orbit around Earth is not a circle — it's an ellipse (exaggerated here), a kind of oval. A supermoon is a new or full Moon that occurs when the Moon is closest to Earth in its orbit. The full supermoon is only slightly larger and brighter than the average full Moon, so if it wasn't in the news, most people wouldn't even notice.

When the Moon is at its farthest from Earth, it's sometimes called a micromoon.

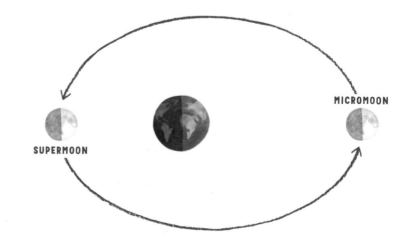

DO A MOONDANCE ★ ★ ★

See if you can dance the way the Moon, Earth, and Sun do in our sky. It's harder than you might think! Which role do you think is easiest?

WHAT YOU NEED:

- 3 people
- An open space, inside or out
- Optional: a sound system

1. Give each person a role: Sun, Moon, or Earth

2. Take your positions:
- ★ The Sun stands in the center of the space.
- ★ The Earth stands to one side of the Sun.
- ★ The Moon stands next to the Earth.

3. Practice your moves:
- ★ The Sun stays in the middle, spinning (rotating).
- ★ The Earth walks around (orbits) the Sun counterclockwise, rotating as it walks.
- ★ The Moon steps around the Earth counterclockwise while always facing it (orbiting and rotating), traveling with the Earth as it orbits the Sun.

4. It's showtime! Perform your Moondance. If you want, put on some moon music, like "Moondance" or "Fly Me to the Moon," and see if you can make 5 big circles around the Sun.

5. Switch roles and try it again!

A day on the Moon lasts about a month in Earth days. That's how long it takes the Moon to rotate, or turn itself around, between one sunrise and the next.

SUN

EARTH

MOON

SIGHTSEEING
on the Moon

The second most noticeable thing about the Moon is that it's covered with craters. Craters form when **meteoroids** (space rocks) crash into the Moon. They look different depending on how large the meteoroid was and how fast it was moving.

Some craters have whitish rays fanning out from their centers, thousands of kilometers long. These rays are made of light-colored rock ejected from the Moon when it was hit by a really big meteoroid.

A **mare** (**mar**-ay) is a round, smooth, dark plain on the Moon. More than one mare are called maria (**mar**-ee-uh). *Mare* means "sea" in Latin: ancient people thought that the maria were oceans on the Moon.

Maria also formed when very large meteoroids crashed into the Moon. These large craters later filled up with lava. We can see them with the naked eye. Use the Moon Map on page 37 to identify them and impress your friends!

Apollo 11
LANDING SIT

1

KEPLER CRATER is noticeable for its prominent rays, created when a meteor struck the Moon's surface and flung debris outward from the impact site.

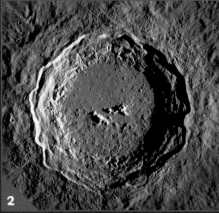

2

COPERNICUS CRATER, less than 1 billion years old, has long clear rays, walls that resemble stair steps, and mountain peaks in its center.

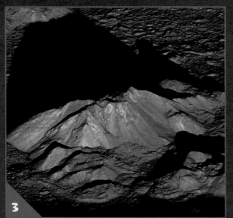

3

TYCHO CRATER, near the bottom of the Moon, is a young and striking lunar crater — white with rays that resemble bright streaks. Its central peak is about 6,500 feet (2,000 m) high.

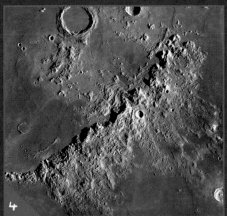

4

APENNINE & CAUCASUS MOUNTAINS are rugged mountain ranges with peaks rising as high as 16,500 feet (5,000 m) from the lunar floor.

The Far Side

We always see the same face of the Moon, because as it orbits, it keeps the same side toward the Earth at all times. The side we see is called the near side, and the side we don't see is called the far side (not the dark side!).

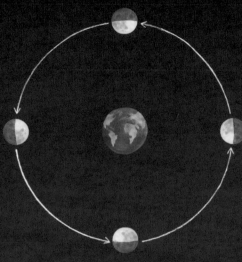

There are more maria on the near side of the Moon than on the far side, because the near-side crust is thinner. Scientists are still not sure why.

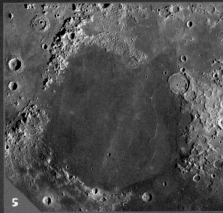

MARE SERENITATIS, or the Sea of Serenity, contains numerous craters and flows of basalt.

MARE TRANQUILLITATAS, or the Sea of Tranquility, has a bluish tint, but not from water — probably from metals in the volcanic rock that covers its surface.

LUNAR LANDINGS

The first spacecraft to carry humans to the Moon was Apollo 11, in July 1969. As Commander Neil Armstrong put his foot onto the lunar surface he said these famous words:

"THAT'S ONE SMALL STEP FOR A MAN, ONE GIANT LEAP FOR MANKIND."

The surface of the Moon is covered in **regolith**, a layer of crushed rock about 15 to 30 feet (5-10 m) thick.

HOW THE MOON FORMED

The most likely explanation for how the Moon was formed is the Giant Impact Model, or "Big Splat." Astronomers think that while our solar system was still forming (see pages 82–83) a Mars-sized planet they named Theia collided with Earth. The Earth was made of molten rock at this time, and most of Theia and its iron **core** sank to the center of the Earth.

In the crash, pieces of the two planets splashed off and formed a disk of debris around Earth. The Big Splat also probably knocked Earth's axis over. Over time, chunks of rock in the disk smashed and melted together, forming larger and larger rocks, until they became the Moon.

After the Moon's **crust** hardened, meteoroids continued to bombard the surface, and some really large ones made huge craters called basins. Erupting volcanoes filled these basins with lava to form the dark-colored maria.

Although volcanic eruptions have ended, meteors continue to make craters on the Moon. There are fewer craters on the maria because those areas have been solid for only about 3 billion years. The surrounding white regions, called the highlands, have 4 billion years' worth of craters.

BASALT

Basalt from volcanoes fills the Moon's maria and makes them look dark.

ANORTHOSITE

The Genesis rock was brought back to Earth by Apollo 15 astronauts. It is made of light-colored anorthosite and is at least 4 billion years old.

Ways the Moon and Earth Are the Same — and Different

★ The Moon and Earth are about the same age.

★ Moon rocks and Earth rocks contain similar ingredients, or compounds.

★ The Moon and Earth are both differentiated. This means that there is denser material at their cores, and less dense material on their surfaces.

★ There is very little water on the Moon.

★ The Moon has more mantle, and less core, than Earth does.

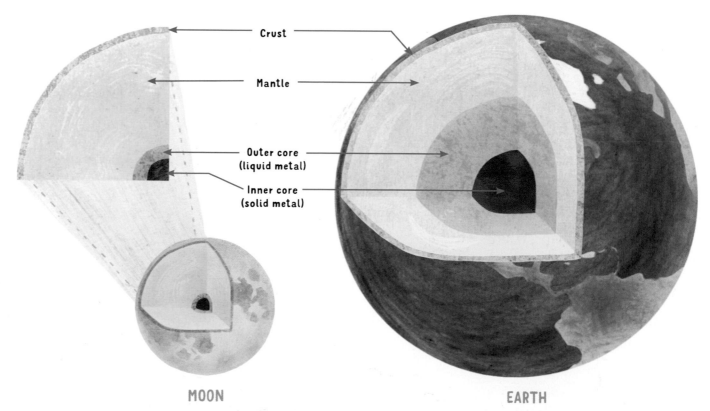

Crust

Mantle

Outer core (liquid metal)

Inner core (solid metal)

MOON

EARTH

THE GIANT IMPACT MODEL

SPLAT!

Blazing bits of molten rock are hurled into space and form a disk circling the Earth.

THEIA

EARTH

4.5 BILLION YEARS AGO

Planet Theia is barreling toward Earth.

Theia crashes into Earth. Its iron core plunges through Earth's molten rock and sinks to the center of the planet.

DURING THE NEXT 100 MILLION YEARS

Chunks of rock in the disk smash together and make larger rocks.

CRUNCH!

BAM!

The rocks mash together and melt into a single orb.

Dense rock and metal sink, and less dense rock floats to the surface.

4.3 TO 3.8 BILLION YEARS AGO

Extra-large meteors bombard the Earth and Moon and make huge craters called basins.

POW!

WHAM!

On the Moon, volcanoes fill these basins with lava and form the maria.

On Earth, weather and volcanic action erase evidence of most craters.

1 TO 2 BILLION YEARS AGO

Volcanic eruptions end, but meteors continue to strike the Moon and make more craters.

PICTURE THE MOON

People see many different things when they look at the Moon. Ancient people made up stories about the shapes they saw on the Moon.

What do you see in the Moon? Make a copy of this picture of the full Moon, and create your own picture with the maria. You can make up a story about it, too!

Man in the Moon
UNITED STATES

Can you see a face in the Moon? Some say the face looks happier in the Southern Hemisphere, sadder in the north. The Moon indeed looks different when you travel: viewed from the Southern Hemisphere the bright Tycho Crater is on top, and the maria form a U-shape, while it's the opposite in the Northern Hemisphere.

Woman in the Moon
NEW ZEALAND

In a Maori legend a woman once mocked the Moon on her way to fetching water. The Moon captured her. You can see the woman in the maria, along with the gourd she used for carrying water.

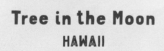

Tree in the Moon
HAWAII

In Polynesian legend the goddess Hina was a gifted weaver who created beautiful kapa cloth from the bark of the banyan tree. She grew restless, though, and left Earth, traveling on a rainbow first to the Sun, which she found too hot, and then to the Moon. There she remained with a banyan tree. You can see the tree, where she lives and continues to weave her cloth.

Moon Rabbit
EAST ASIA

Many cultures around the world see a rabbit in the shapes on the Moon. In Chinese folklore it is using a mortar and pestle to grind herbal medicines for the gods.

Handprints
INDIA

A myth from India says that the Earth goddess placed her hands on the cheeks of her daughter, the Moon, to say goodbye, and the prints remained there.

MOON MAP

Maria and large craters on the moon that you can see with the naked eye or binoculars.

1 **Copernicus**
(CRATER)

2 **Mare Imbrium**
(SEA OF RAIN)

3 **Mare Serenitatis**
(SEA OF SERENITY)

4 **Mare Tranquillitatus**
(SEA OF TRANQUILITY)

5 **Mare Crisium**
(SEA OF CRISES)

6 **Mare Fecunditatis**
(SEA OF FERTILITY)

7 **Mare Nectaris**
(SEA OF NECTAR)

8 **Tycho**
(CRATER)

9 **Mare Nubium**
(SEA OF CLOUDS)

10 **Mare Humorum**
(SEA OF MOISTURE)

11 **Oceanus Procellarum**
(OCEAN OF STORMS)

LUNAR ECLIPSE

The Sun shines on Earth and sends our planet's shadow out into space. A **lunar eclipse** happens when the full Moon passes into the Earth's shadow. But the Moon's orbit around the Earth is tilted compared to the Earth's orbit around the Sun, so the Sun, Moon, and Earth can't line up perfectly for a lunar eclipse every full Moon. There are only about two lunar eclipses per year. (And there are two **solar eclipses**; more on that in the next chapter!)

The shadow that Earth casts on the Moon has two parts. The outer edge of the shadow, called the **penumbra**, is dim and fuzzy. The dark center of the shadow is called the **umbra**.

If the Earth's umbra doesn't completely cover the Moon, we have a partial lunar eclipse. A penumbral lunar eclipse happens when only the penumbra covers all or part of the Moon.

If you look at your own shadow on a sunny day, you can see that it has a fuzzy edge. This is *your* penumbra!

When three objects in the solar system line up, it's called a **syzygy** ("siz-a-gee"). (Use this word the next time you play Hangman!)

Some people call the red-tinged lunar eclipse a blood moon.

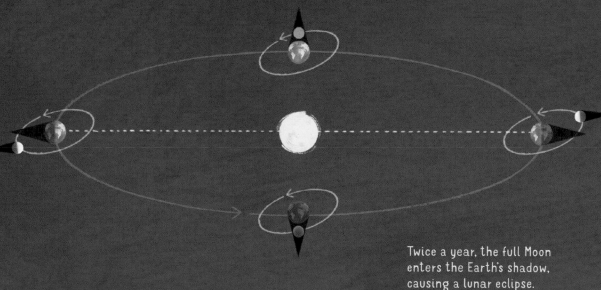

Twice a year, the full Moon enters the Earth's shadow, causing a lunar eclipse.

Stages of a Lunar Eclipse

P1 (FIRST CONTACT).
The Moon first moves into the penumbra. Beginning of the penumbral eclipse.

U1 (FIRST CONTACT).
The Moon first moves into the umbra. The full Moon will look like it has a bite taken out of it. Beginning of the partial lunar eclipse.

U2 (SECOND CONTACT).
The Moon is completely in the umbra. Beginning of **totality** (total eclipse).

MID-ECLIPSE. Halfway through totality. The darkest part of the eclipse.

U3 (THIRD CONTACT).
The Moon starts moving out of the umbra. End of totality.

U4 (FOURTH CONTACT). The Moon moves fully out of the umbra (but is still in the penumbra). End of the partial lunar eclipse.

P4 (FOURTH CONTACT). The Moon moves out of the penumbra. End of the penumbral eclipse.

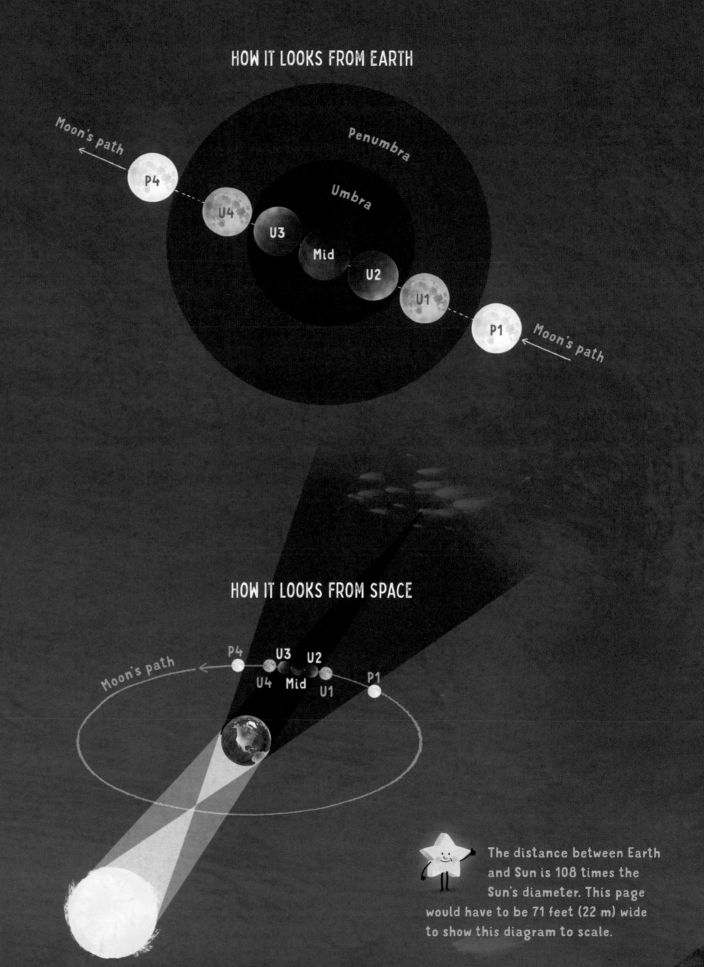

HOW IT LOOKS FROM EARTH

Moon's path

Penumbra

Umbra

P4 · U4 · U3 · Mid · U2 · U1 · P1

Moon's path

HOW IT LOOKS FROM SPACE

Moon's path

P4 · U3 · U2 · U1 · P1
U4 · Mid

The distance between Earth and Sun is 108 times the Sun's diameter. This page would have to be 71 feet (22 m) wide to show this diagram to scale.

How to Observe a Lunar Eclipse

It is safe to observe a lunar eclipse with the naked eye (unlike a solar eclipse).

WHEN: Look up the times for the eclipse contacts online. Look carefully! Many eclipse websites will give you times in UTC (Coordinated Universal Time), which is the time in Greenwich, England. Be sure to find your location and your local time.

WHERE: Find a place where you can see the Moon during the eclipse. The full Moon is near the eastern horizon around 6:00 p.m., high in the southern sky at midnight (northern sky in the Southern Hemisphere), and near the western horizon around 6:00 a.m.

HOW: Sit in a beach chair or lie down on a blanket so your neck doesn't get tired.

If you have binoculars, look through them to observe the Moon closely.

 About 2,000 years ago, the ancient Greeks realized that the circular shape of Earth's shadow during a lunar eclipse means that the Earth is a sphere.

BRING BACK THE MOON!

In earlier times, people of many different cultures thought that eclipses happened when some kind of creature gobbled up the Moon. They would make a lot of noise and shoot guns or arrows into the air to scare the creature away. Even today, at the Griffith Observatory in Los Angeles, eclipse watchers celebrate by playing instruments and banging on pots and pans. Naturally, the Moon comes back every time!

The Moon will come back, I promise!

The **terminator** is where, if you were standing on the Moon, you would see the Sun set.

The Moon

Binoculars will show you more detail on the Moon than you can see with your naked eye.

The boundary between light and dark on the Moon is called the terminator. Shadows are longer along that line, and you can see craters more clearly. When the Moon is full, it has no terminator. The full moon looks like a bright, flat disk, and you can see only the maria clearly.

During the Moon's crescent phase, you may be able to see earthshine through binoculars (see page 28), even if you can't see it with the naked eye. Binoculars may also show you craters and maria in the dark regions.

When you observe the Moon with binoculars, take along your red flashlight, the Moon Map (page 37), and your Astronomy Notebook. Drawing what you see through the binoculars can help you notice more detail. The longer you look, the more you'll see!

Lunar eclipses are fun to watch through binoculars, too. You can watch the Earth's shadow as it moves across the Moon's face. During the total eclipse phase, when the Moon is dimly lit with a faint red glow, it looks like the sphere it is, not a disk.

the SUN

The Sun is the center of our solar system and its most important object. Its gravity keeps all the planets in their orbits. Its light provides energy for all life on Earth. For millennia, humans have tracked the Sun in order to tell time and predict the seasons.

SUNWISE

The Sun is easy to observe. It's big and bright, it's up during the daytime, and you can see it even if the sky is partly cloudy. As with the Moon, light pollution is never a problem!

However, you must NEVER look directly at the Sun with your naked eye. Your eye's lens focuses the Sun's light on the retina, which can cause permanent damage. This chapter explains how to observe the Sun safely.

Tracking the Sun's Path

The Sun takes a slightly different path through the sky every day: it rises, transits, and sets in different places. You can test this by keeping track of where the Sun sets on your horizon. Even after only a week or two, you will see that the point where the sun sets changes.

Stonehenge is an ancient stone circle in England. Scientists believe that the people who lived there were farmers who put up stones to mark the rising and setting points of the Sun. It would have been important for farmers to keep track of the seasons in order to plan ahead. Special times of the year were marked by special stones.

At Stonehenge, built thousands of years ago, the Sun sets (and rises) near different stones at different times of the year, as seen from the center of the circle.

Medicine Wheel/Medicine Mountain, a National Historic Landmark located in Wyoming's Big Horn Mountains, has been marking time through the seasons for hundreds of years.

Saturn

Cityhenge

If your neighborhood is laid out on a grid, you may be able to experience Cityhenge. That's the one or two days of the year when the setting Sun is visible between buildings on all the streets of your city. For Manhattan in New York City, this occurs twice: at the end of May and in mid-July. (You can search online for the exact dates.)

Is your neighborhood a grid? Check an online map. If so, watch to find out which days the rays of sunset shine directly down your street.

ASTRONOMY NOTEBOOK

MAKE A SUNSET CALENDAR ★ ★

You can make a sunset calendar in your neighborhood. Here's how. (Note: This project will take a whole year!)

1. Find a place where you can see the western horizon and that you can easily visit all year. This will be your observing point.

2. Check a map or satellite picture online to figure out what direction is west from your observing point. Don't use a compass! (We'll explain why not later in this chapter.)

3. Make a sketch of your western horizon.

4. Check online to see when the Sun will set. Go to your observing point about half an hour before sunset. For astronomers, sunset is the time when the Sun crosses the horizon. If there are buildings or hills on your horizon, the Sun will "set" a little earlier than the official time.

5. Wait for the Sun to set. When it does, put an X on your horizon picture at the point where it set, and label the X with the date and time.

6. Repeat step 5 about once a week. Make sure to observe the sunset on special days, like the first day of each season, holidays, and your birthday.

After a full year of observations, you will know where the Sun sets in your neighborhood at different times of the year. You can use the picture of your horizon as a calendar.

If you're an early bird, you can do this experiment with sunrises. They will follow the same pattern as the sunsets, but on the eastern horizon.

HOW SEASONS HAPPEN

The Earth circles the Sun in a path called an **orbit**. It takes a year to go all the way around the Sun in one complete orbit and one seasonal cycle.

The Earth's **axis of rotation** passes through its north and south poles. Earth's North Pole always points toward Polaris, the North Star. But Polaris is not straight above the Earth's orbit. The Earth's rotation axis is tilted from its orbit around the Sun by 23.5 degrees.

As the Earth goes around the Sun, the Sun appears to be north or south of our **equator**, depending on the season — where Earth is in its orbit.

★ **ON MARCH 21 AND SEPTEMBER 21** (give or take a day), the Sun is directly over the equator. These days are called **equinoxes**, and they are the first days of spring and fall.

★ **ON JUNE 21,** the first day of summer in the Northern Hemisphere, the Sun reaches its northernmost point, 23.5 degrees north of the equator.

★ **ON DECEMBER 21,** the first day of winter in the Northern Hemisphere, the Sun reaches its southernmost point, 23.5 degrees south of the equator. These days are called **solstices**.

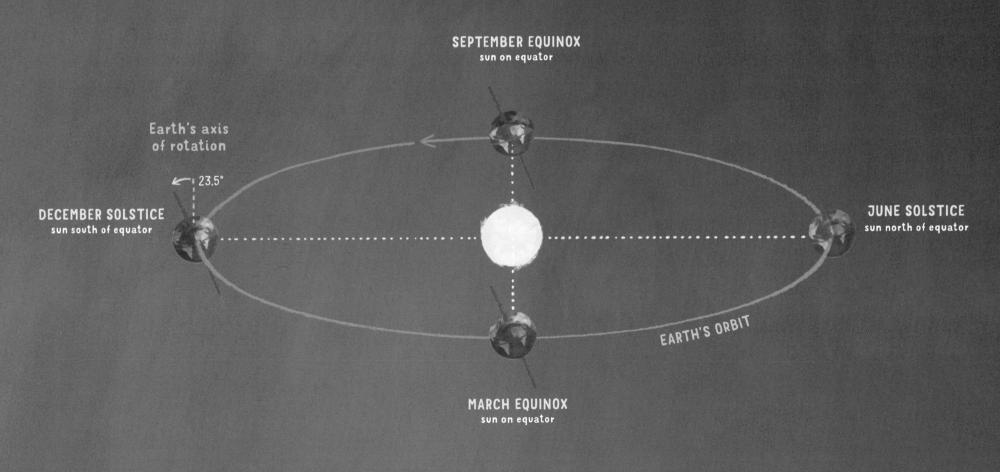

SEPTEMBER EQUINOX
sun on equator

Earth's axis
of rotation

23.5°

DECEMBER SOLSTICE
sun south of equator

JUNE SOLSTICE
sun north of equator

EARTH'S ORBIT

MARCH EQUINOX
sun on equator

HOW SEASONS LOOK FROM SPACE

SUN PATHS

The Sun's path that we observe is really just part of a complete circle that the Sun seems to make in the sky during a 24-hour day. The fraction of the circle above the horizon tells you how long the Sun will be up. The fraction below the horizon tells you how long night will be.

On the equinoxes in March and September, the Sun rises due east and sets due west. Daylight lasts for 12 hours.

On the December solstice, the Sun rises in the southeast and sets in the southwest. In the Northern Hemisphere, the day is short and the night is long. The Sun is low in the sky, and it's winter.

On the June solstice, the Sun rises northeast and sets northwest. In the Northern Hemisphere, the day is long and the night is short. The Sun is high in the sky, and it's summer.

The Southern Hemisphere has the opposite seasons from the Northern Hemisphere. The Sun is low in the sky in June and high in the sky in December.

HOW SEASONS LOOK FROM EARTH
Sun's path in the sky at 40° North

Moon Paths

The full Moon is on the opposite side of the sky from the Sun. In winter, the full Moon is high in the sky and above the horizon for more than 12 hours. In the summer, the full Moon's path is low in the sky.

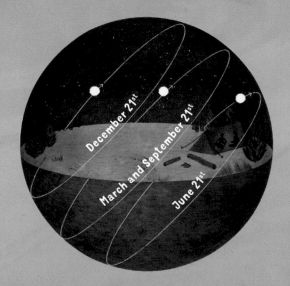

In the Arctic and Antarctic circles, there are some days when the Sun never sets. This is a timelapse photo of the "midnight sun," taken every hour for 20 hours on the summer solstice.

The Sun's Path in Different Parts of the World

The path of the Sun looks different depending on how far you are from the equator — your latitude. If you're near the equator, the Sun's path is high in the sky. Far from the equator, the Sun's path is low in the sky. If you're near the North or South Pole, it is really low. These differences affect the climate in different places on Earth.

Right at the North and South Poles there are six months of daylight in the summer and six months of darkness in the winter. The Sun rises and sets on the equinoxes.

SUN'S PATH AT THE EQUATOR

SUN'S PATH AT THE NORTH POLE

TELLING TIME BY THE SUN

Ancient people kept track of time during the day by noticing the Sun as it rose, made its way across the sky, and set. The Sun reaches its highest point when it transits, or crosses, the meridian. This is halfway through the day, called noon. The time before noon is called a.m., for the Latin words *ante meridiem* ("before the meridian"). Afternoon is p.m., for *post meridiem* ("after the meridian").

The First Clocks

The first clocks were sundials. The time from sunrise to sunset was divided into twelve equal parts, called "hours." The first hour began at sunrise, the sixth hour began at noon, and the twelfth hour ended at sunset.

Because the Sun is up longer in the summer than in winter, summer hours were longer than winter hours on a sundial. Around 1300 C.E., people started using mechanical clocks that marked twenty-four equal hours in a day.

The simplest sundial is just a stick in the ground, like the one we'll make on page 49. This doesn't make a very good clock, though, because the Sun's path — and the shadow's path — is different every day. People have created much fancier sundials in all kinds of shapes and sizes that will work every day of the year.

All sundials have a **gnomon** (pronounced no-min) and a *dial*. The gnomon casts a shadow on the dial, and the shadow moves throughout the day as the Sun crosses the sky. The gnomon's shadow is like the hands on a clock, and the dial is like the clock's face.

 In the Northern Hemisphere, a gnomon's shadow moves clockwise around the dial. (That's why clock hands move in that direction.) *Sunwise* is an old word that means clockwise.

Time Zones

Time on Earth depends on where the Sun is in the sky, which is different depending on where you are on Earth. In this picture, it's noon in Nairobi, because the Sun is directly overhead. In Singapore, it's 5:00 p.m. And in Quito, Ecuador, it's 4:00 a.m., and the Sun hasn't risen yet!

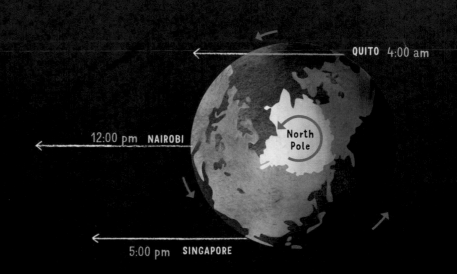

QUITO 4:00 am

12:00 pm NAIROBI

North Pole

5:00 pm SINGAPORE

TRACK THE SUN ✦

You can track the Sun in the sky without looking directly at it by using a stick to cast shadows on the ground. Do this on a bright, sunny day when you can visit your sundial throughout the day.

YOU WILL NEED:

- Place with a good view of the south
- Fairly straight stick
- Modeling clay (optional)
- Several pebbles or sidewalk chalk

1. Early in the morning, go to the place you've chosen for your sundial and push the stick into the ground so that it stands straight up. If you're on a roof or pavement, use modeling clay to stand the stick up.

2. The Sun will cast a shadow of the stick on the ground. Mark the end of the shadow with a pebble or chalk.

3. Mark the end of the shadow every hour or so.

The stick's shadow tells you where the Sun is in the sky. When the Sun is high in the sky, the shadow is short; when it's low, the shadow is long. The shadow points in the opposite direction from the Sun.

You can try this activity at different times of year. The shadows will be different because the Sun's path is different, but the shortest shadow will always point north.

Finding North by the Sun

Every 10 minutes between 11 am and 1 pm (noon and 2 pm during Daylight Saving Time), mark the end of the stick's shadow. Then find the mark that shows the shortest shadow. Draw a line between it and the stick. This line points due north!

To track the Sun in the Southern Hemisphere, just replace the word "north" with "south" in these instructions. In the Southern Hemisphere, the Sun reaches its highest point in the northern sky, and its shadow points due south.

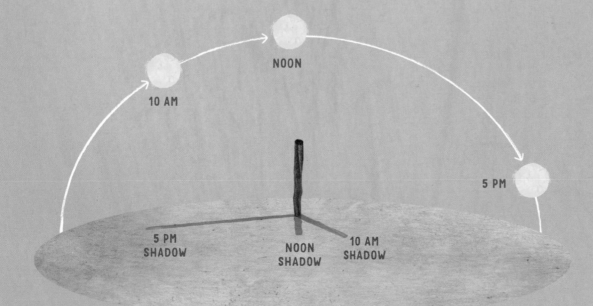

NOON

10 AM

5 PM

5 PM SHADOW

NOON SHADOW

10 AM SHADOW

A Visit To THE SUN

We can learn a lot about stars by studying the Sun, the closest and brightest star in our sky.

CHROMOSPHERE
The Sun's upper atmosphere. Only visible during an eclipse or through special telescope filters

CORE
Interior of the Sun, where fusion reactions occur

CORONA
Thin, outermost layer of the Sun's atmosphere. Only visible during an eclipse or with a special telescope

PHOTOSPHERE
The everyday surface of the Sun

PROMINENCES
Giant loops of gas

SUNSPOTS
Cool areas of the photosphere (page 51)

EARTH

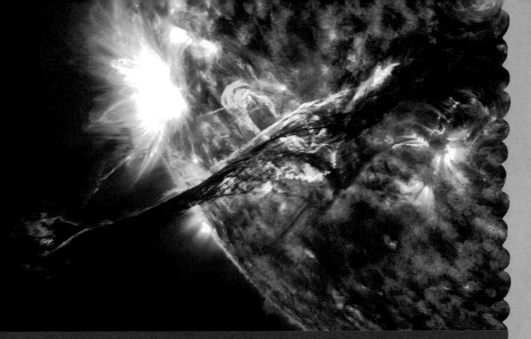

STORMS ON THE SUN

A sunspot is a dark patch that appears from time to time on the Sun. It occurs where a magnetic storm has trapped gases on the surface, causing that spot to cool down. Sunspots can be up to 10 times the size of the Earth and last for a few days to a month.

Sunspots appear to move across the Sun because the Sun rotates, just like the Earth. By observing sunspots, astronomers have discovered that the Sun's equator rotates faster than its poles!

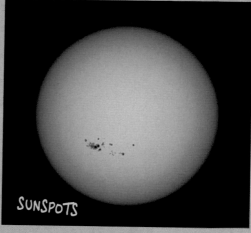

SUNSPOTS

WHEN CAN WE SEE THEM?

The Sun has an 11-year cycle of sunspots and magnetic activity. During the period of greatest activity, called solar maximum, there are dozens of sunspots at a time. During solar minimum, which happens 5½ years later, you can go for days without seeing a single spot.

AURORA TIME!

Solar maximum is also the best time to see the aurora, because the solar wind is stronger and there are more coronal mass ejections. These clumps of gas explode off the surface of the Sun and travel out with the solar wind.

Powerhouse

The Sun is 73 percent hydrogen, 25 percent helium gas, and 2 percent everything else. All the atoms in the Sun are **ions** — every electron is free from its nucleus.

In the core of the Sun, hydrogen nuclei smash into each other at high speeds. In a multi-step process, six hydrogen nuclei fuse together to make one helium and two hydrogen nuclei, plus a little bit of energy (light and heat). This is called nuclear **fusion**. Fusion energy makes the Sun and other stars shine.

Hot Stuff

The Sun's diameter is 864,000 miles (1,391,000 km). That's 109 times as big as Earth! And its **mass** — the amount of matter it contains — is 333,000 times

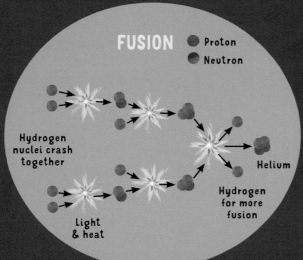

FUSION   ● Proton
● Neutron

Hydrogen
nuclei crash
together

Helium

Light
& heat

Hydrogen
for more
fusion

the mass of Earth, or 2×10^{30} kg (that's a two followed by thirty zeros). The sun has enough hydrogen fuel to last 10 billion years.

The core of the Sun is incredibly hot: 27,000,000°F (15,000,000°C). At the Sun's surface, or photosphere, it's only 9,900°F (5,500°C).

There are areas on the photosphere called **sunspots** that are even cooler, at around 7,000°F (4,000°C).

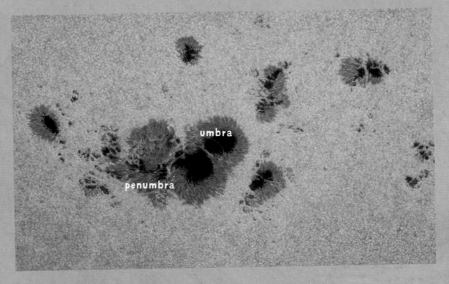

umbra

penumbra

Sunspots have a very complex structure, as you can see from this close-up. Every sunspot is unique.

HOW THE SUN FORMED

Our Sun, like all stars, was born in a nebula, a massive, spinning cloud of gas (mostly hydrogen) and dust. Inside a nebula are denser clumps of gas. The force of **gravity** makes these clumps collapse into **protostars** (early stars).

As a protostar grows denser and more massive, it starts to heat up and glow. When its core gets hot enough, **nuclear fusion** begins to turn the hydrogen into helium. The protostar has become a star!

Stars that are powered by hydrogen fusion are called **main sequence stars**. Our Sun is one. Stars spend most of their lives in this phase.

A main sequence star is stable. Gravity tries to crush it inward, while the fusion reactions inside try to blow it up. These two forces perfectly balance each other out.

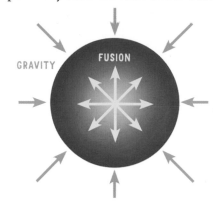

As the proto-Sun formed, the spinning gas cloud flattened, flinging material out into a disk surrounding the Sun. That material would eventually become planets (see page 82).

Where stars are born: the Eagle Nebula, one of the most famous star formation regions, is nicknamed the "Pillars of Creation."

Fusion on Earth

Scientists are working to make power plants that use fusion, an energy source that doesn't produce greenhouse gases. Instead of gravity, they use super-strong magnetic fields to squeeze a mass of hydrogen gas down to the density found at the center of the Sun. The designers of ITER, a huge nuclear fusion plant that's under construction in France, hope it will able to produce fusion power on a large scale by 2035.

Nebulae come in all shapes and sizes. This is the Orion Nebula.

OUR SUN Is BORN

4.5 BILLION YEARS AGO

A giant, spinning cloud of gas and dust forms, the future birthplace of our Sun.

SWOOSH!

As the cloud spins, gravity forces it to collapse inward into a disk, still spinning.

It is now a PROTOSTAR, or early star.

The dense center of the disk heats up as it collapses.

This heat shoves out against gravity and stops the collapse.

Our Sun is born.

AFTER A FEW MILLION YEARS

Over time, the materials spinning around the Sun cluster together into rocks, then PLANETESIMALS, and finally PLANETS. We'll talk about them in chapter 4.

POW!

CRACK!

ZOOOOM!

SOLAR ECLIPSE

A solar eclipse happens when the Moon comes between the Sun and the Earth and completely blocks out the Sun.

During a solar eclipse, the shadow of the Moon falls onto the Earth. The dark part of the shadow is called the umbra; the soft blurry part between the dark and light areas is the penumbra. Everyone in the umbra sees a total eclipse; everyone in the penumbra sees a partial eclipse. Anyone outside the path of the Moon's shadow won't see an eclipse at all!

A total solar eclipse happens somewhere on Earth about every year and a half. Any given place on Earth will have a total solar eclipse every 375 years on average.

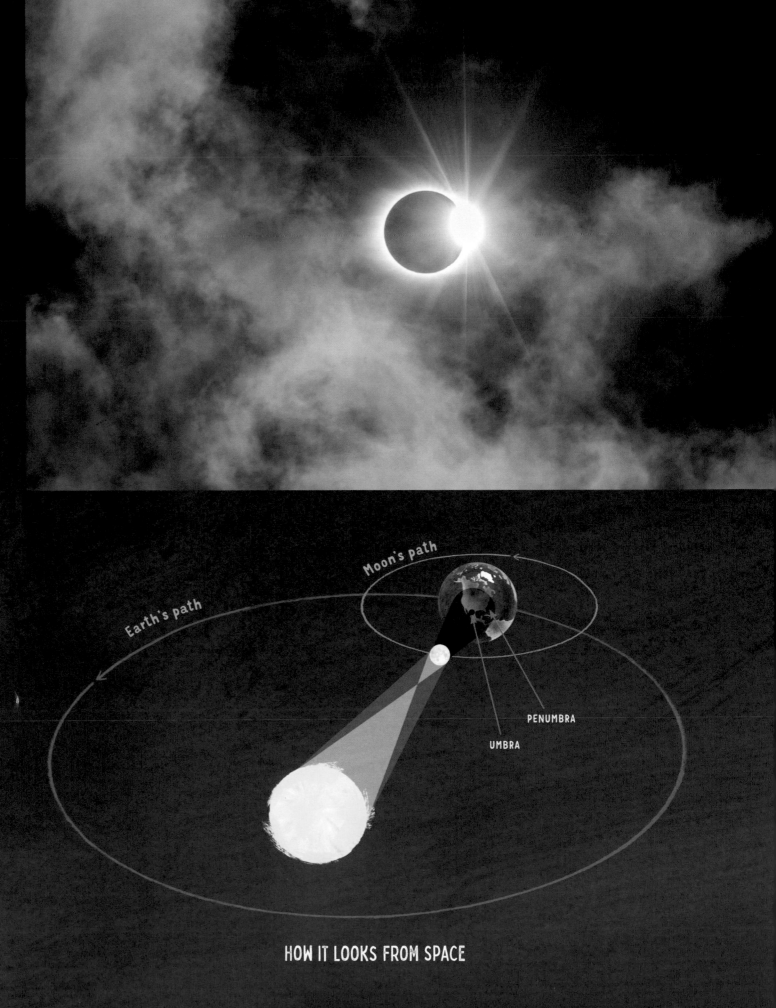

Moon's path

Earth's path

PENUMBRA

UMBRA

HOW IT LOOKS FROM SPACE

Stages of a Solar Eclipse

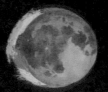

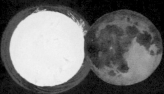

First contact   Second contact   Totality   Third contact   Fourth contact

·········· eclipse glasses on ··········   ·········· eclipse glasses on ··········

FIRST CONTACT. The Moon touches the edge of the Sun; partial solar eclipse begins.

What can you see?

★ Viewed through eclipse glasses, the Sun will look like it has a tiny bite taken out of it.

★ It gets darker and colder as the Moon covers more of the Sun.

★ You may see *shadow bands* a few minutes before and after totality (total eclipse). These are wiggly stripes of light and dark that move quickly across the ground. They're caused by the light of the thin crescent Sun coming through the Earth's wiggly atmosphere. They are easiest to see on a white sheet, wall, or poster board.

★ If you have a clear view of the horizon, you can see the Moon's shadow rushing across the Earth toward you before the eclipse and away from you after the eclipse.

SECOND CONTACT. Beginning of totality: the Moon covers the Sun completely. ***When you can't see the Sun through your eclipse glasses, it's safe to take them off for this period only — and be ready to put them on again soon.***

Baily's beads

Diamond Ring effect

Corona

What can you see?

★ Just before totality, the *diamond ring effect* and Baily's beads — the last bits of sunlight shining between the mountain peaks on the Moon.

★ The ghostly *corona* (see page 50) around the black disk of the Moon. It looks different every eclipse.

★ The Sun's upper atmosphere, called the *chromosphere*, will be a faint red ring around the Moon.

★ *Prominences* — giant arcs of gas on the Sun's surface — that look like red loops sticking out of the edge of the Moon.

★ Bright stars and planets in a dark sky.

★ A "sunset" all the way around the horizon (because it's dark where you are and light everywhere else).

★ Strange animal behavior, such as birds growing quiet. During totality, animals think it's nighttime!

THIRD CONTACT. End of totality. ***Put your eclipse glasses back on to protect your eyes.***

What can you see?

★ The edge of the Sun starts to show again. The eclipsed Sun will suddenly be bright again.

★ Baily's Beads and the Diamond Ring appear again, this time on the other edge of the Sun.

★ Watch the crescent Sun grow as the Moon moves away.

FOURTH CONTACT. End of partial solar eclipse.

What can you see?

★ The entire Sun is visible.

★ Birds and animals will behave normally again.

How Are Eclipses of the Sun and Moon the Same — and Different?

Solar eclipses happen during the new Moon, but like lunar eclipses, they don't happen every month. Solar and lunar eclipses usually occur two weeks apart from each other.

Shadow Play

The shadow of the Moon on Earth during a solar eclipse is much smaller than the huge shadow we cast on the Moon during a lunar eclipse. During a lunar eclipse, the entire Moon is in shadow. Anyone on Earth who can see the Moon will see the lunar eclipse, which can last for more than an hour.

During a solar eclipse, only a tiny part of Earth is in shadow. Only people directly on the path of this shadow will see totality. And because the shadow is moving very fast, they will see it for only 7 minutes at the most.

Weather satellites captured this image of the umbra and penumbra during the August 2017 total solar eclipse.

Why Can't We Look at the Sun?

Looking at the Sun is dangerous because the retina in your eyes doesn't have any nerves that feel pain, so you can damage your eyes without knowing it. Many people believe that the Sun's rays are even more dangerous during an eclipse. In fact, they are very dangerous for our eyes at all times — but it's much more tempting to look at the Sun during an eclipse. **Never look directly at the Sun, whether there's an eclipse or not.**

OBSERVING THE SUN SAFELY

The only safe way to look at the Sun is to wear eclipse glasses. They have special filters that block out 99.999 percent of the Sun's light. When you look through them, you should not be able to see anything but the Sun.

It is *not* safe to view the Sun through things like potato chip bags, DVDs, or smoked glass, no matter what you read on the Internet.

See Resources for companies that sell good-quality eclipse glasses. Before using them, hold your eclipse glasses up to a bright indoor light to look for scratches and holes. If you find any, cut up the glasses and throw them out.

A Moving Target

Unlike solar eclipses, which are visible only on a small part of the Earth, lunar eclipses can be seen by anyone who's on the same side of the Earth as the Moon.

Since the Moon is sometimes closer to us and sometimes farther away (see page 30), it often appears too small to completely block out the Sun. At those times we see an **annular eclipse**.

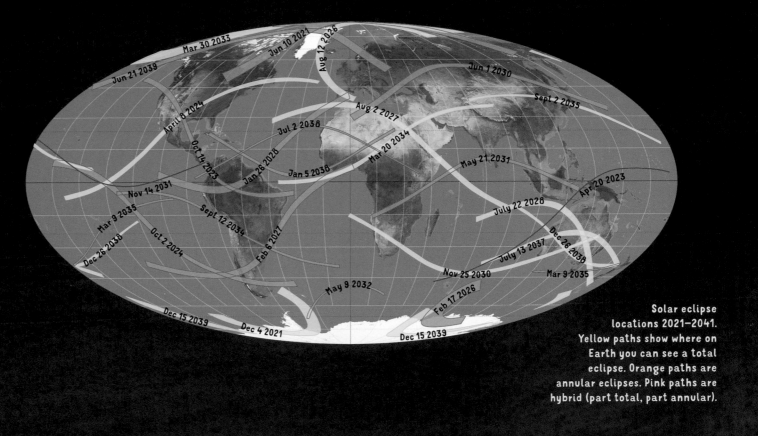

Mar 30 2033
Jun 10 2021
Aug 12 2026
Jun 1 2030
Jun 21 2039
Sept 2 2035
April 8 2024
Aug 2 2027
Jul 2 2038
Oct 14 2023
Mar 20 2034
Jan 26 2028
May 21 2031
Jan 5 2038
Apr 20 2023
Nov 14 2031
July 22 2028
Mar 9 2035
Dec 26 2038
Sept 12 2034
Feb 6 2027
July 13 2037
Dec 26 2038
Oct 2 2024
Nov 25 2030
Mar 9 2035
May 9 2032
Feb 17 2026
Dec 15 2039
Dec 4 2021
Dec 15 2039

Solar eclipse locations 2021–2041. Yellow paths show where on Earth you can see a total eclipse. Orange paths are annular eclipses. Pink paths are hybrid (part total, part annular).

WHAT TO TAKE ECLIPSE-WATCHING

Solar eclipses are exciting because they're rare, and they're an opportunity for people all over the world to observe together. If this is your first eclipse, don't take too much fancy equipment. You should spend most of your time watching the eclipse, not trying to get your camera or telescope to work.

Here's what you *must* have:

★ Eclipse glasses (see Resources)

Other useful items:

★ A pinhole projector (see page 58)
★ A white sheet or piece of poster board to see shadow bands (see page 55)
★ Snacks and water
★ Sunblock and hat
★ Watch or phone
★ List of eclipse contact times
★ Astronomy Notebook and pencil
★ Camera
★ Binoculars with solar filters

Never thought I'd see a crescent Sun.

MAKE A PINHOLE PROJECTOR ★★

One way to observe an eclipse is to use a pinhole projector to make an image of it.

YOU WILL NEED:

- Scissors
- Sheet of plain white paper
- Tape
- Shoebox with lid
- Aluminum foil
- Pin or thumbtack

white paper

aluminum foil

shoebox

hole made with pushpin

cut-out door

1. Cut out a rectangle of white paper and tape it to the inside of one end of the box.

2. On the other end of the box, cut out small square that is about 1 inch by 1 inch (2.5 cm by 2.5 cm).

3. Cut a piece of foil that is 2 inches by 2 inches (5 cm by 5 cm). Poke a tiny hole in the center of the foil with the pin.

4. Tape the foil over the square hole in the box.

5. Cut a door in the side of the box so that you can peek in and see the white paper screen.

6. Put the lid on.

To use the projector, stand with your back to the Sun and put the box on your shoulder, with the pinhole facing the Sun. Don't look through the pinhole at the Sun! Peek in the box through the door. Turn the box until you can see an image of the Sun on the white paper screen.

You can use any box for this project, or a poster tube. The longer the box is, the larger the image of the Sun will be. Make sure your box is nice and dark on the inside. You can use your pinhole projector to view a solar eclipse or even very large sunspots. The image will be too small to see fine details on the Sun.

Have you ever noticed that when you sit under a shady tree, there are little circles of sunlight on the ground among the leaf shadows? If you look under a leafy tree during an eclipse, you'll notice that those circles are now crescent shaped! The spaces between the leaves act as pinholes and form images of the eclipsed Sun on the ground.

A Simple Pinhole Viewer

A much simpler pinhole projector can be made from two paper plates. Poke a tiny hole in one plate, and hold it above the other. The pinhole will make an image of the Sun on the bottom plate.

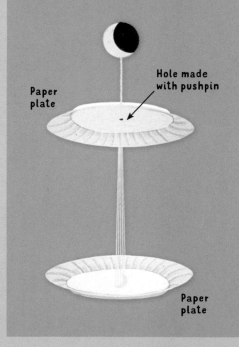

Hole made with pushpin

Paper plate

Paper plate

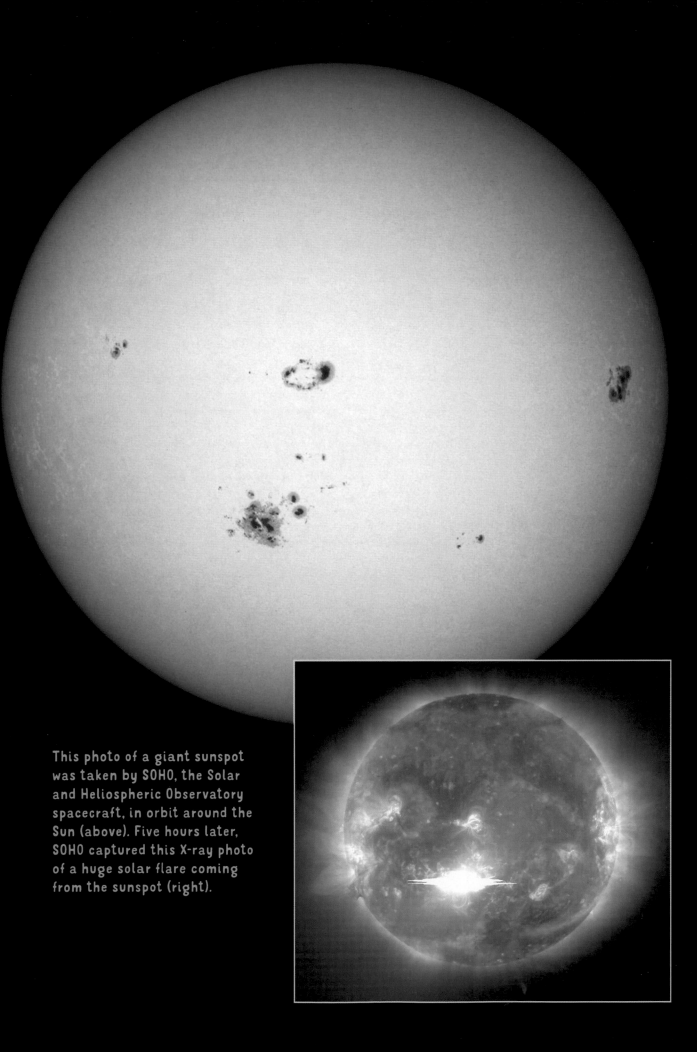

This photo of a giant sunspot was taken by SOHO, the Solar and Heliospheric Observatory spacecraft, in orbit around the Sun (above). Five hours later, SOHO captured this X-ray photo of a huge solar flare coming from the sunspot (right).

Our Sun

Binoculars let in more light than your eyes do, so it's even more dangerous to look at the Sun through binoculars than it is to look at it with your naked eye.

You can buy solar filters at the same places you buy eclipse glasses (see Resources). Make sure they are the right size to fit on your binoculars without falling off. And make sure to buy two! Be sure that the filters you buy are approved for solar viewing. Do not try to make your own filters.

Solar filters should always go on the objective lenses of the binoculars (see page 24), not the eyepieces, otherwise the Sun's light could damage the lenses inside the binoculars and your eyes. Before you put solar filters on the binoculars, check them for holes and scratches, just as you would check eclipse glasses.

Binoculars with solar filters can also help you look for sunspots. Check an online image of the Sun before you go out observing so that you know what sunspots you should be able to see. Make a sketch of the Sun every few days in your Astronomy Notebook and watch the sunspots move across the Sun's face.

PLANETS

Our Earth is part of a solar system. Our Solar System includes planets, asteroids, comets, and dust, with the Sun at its center keeping everything in orbit. We now know that most stars have planets. But we know of only one planet that has life on it: our home, Earth.

EARTH'S SIBLINGS IN THE SKY

Planets shine by reflecting the light of the Sun, unlike stars, which shine with their own light. We can see five of the planets in our solar system with the naked eye: Mercury, Venus, Mars, Jupiter, and Saturn.

The planets rise and set, just like the Sun, Moon, and stars do. As the Earth and planets move around the Sun, the planets change position in our sky.

Like the Moon, the planets don't rise and set at the same time every day. Each planet has its own

The soft cluster at left is the Pleiades — but to Northern Hemisphere viewers it looks upside-down here, because this photo was taken from Earth's Southern Hemisphere.

orbit and its own motion in the sky. They all stay near the ecliptic (see page 62) and usually rise a bit earlier each day. How much earlier depends on how far they are from Earth and where each planet is in its orbit around the Sun.

Pleiades

Jupiter

Venus

Aldebaran

 The solar system is a big place: it takes light (the fastest thing possible) 8 minutes to get from the Sun to the Earth, 42 minutes to reach Jupiter, up to 7 hours to arrive at Pluto, and more than a year to reach the most distant comets.

STAR OR PLANET?

To an observer on Earth, the planets can look like bright stars. Here's how to tell planets and stars apart.

Where Is It?

The **stars** appear everywhere in the sky. They all "travel" east to west (in reality, our Earth is turning west to east), but they take many different paths.

Planets, on the other hand, are located only on the **ecliptic**, a line in the sky that marks our solar system. The Sun and Moon are on that line, too — which is how they "meet" in an eclipse!

The planets in the solar system (including Earth) all orbit the Sun in almost the same **plane**, or the same flat surface. The Moon circles the Earth on that plane as well.

The Moon, the Sun, and You

To visualize this plane, stand outside and look for three objects in the solar system. The Moon is one object. Its bright side points to the Sun, which is the second object. You, standing on Earth, are the third object.

Now imagine that the Earth, Moon, and Sun are all sitting on the same plane, like marbles on a plate. All of the planets are also on this plate. The ecliptic is an imaginary line that maps out the edge of the plate all the way around you.

If you are in the **Northern Hemisphere**, then all of the planets travel across the southern half of the sky, like the Sun and Moon do. In the **Southern Hemisphere**, you will find them in the northern part of the sky.

How Bright Is It?

Planets are some of the brightest objects in the night sky.

Is It Colorful?

Mars is red, and Saturn is yellowish. (Unfortunately, the other planets look white!)

Jupiter Moon Ecliptic Sun Ecliptic

This is an easy way to find the ecliptic. The Sun, Moon, and you are all on the same plane, like a flat plate connecting the solar system.

Does It Twinkle?

Stars twinkle because their light comes to us through Earth's atmosphere. The atmosphere bends a star's light toward or away from our eyes, making the star seem to jump around or change brightness. The star also appears to be in a place that is far away from its actual position.

Stars are so far away that they are tiny points of light. Planets are much closer, so they appear larger than stars. When a planet's light comes through the atmosphere, it also is bent, but most of it still gets through

to our eyes, so we don't notice the twinkling. When a planet is close to the horizon, its light has to pass through more of the atmosphere to reach our eyes, and then it can seem to twinkle.

Look It Up!

If you see an object that you think is a planet, look at a star chart (or one of the seasonal sky maps in chapter 5). If that bright thing is not in a constellation, it's probably a planet. Planets seem to travel from one constellation to another in their trip around the Sun, but stars stay in their constellations.

Any bright, non-twinkling "star" near the ecliptic is probably a planet. The ecliptic is marked on the star maps in this book, so you can also find it among the constellations.

Star's actual position

Star's apparent position

Planet

Atmosphere

Jupiter

PLANETS INSIDE AND OUTSIDE

Venus and Mercury are between us and the Sun. They are called **inferior planets**. In our sky, they appear near the Sun, leading the Sun up at dawn or following it down at dusk. They are visible for only a few hours.

Venus and Mercury are the only planets that transit, or pass between Earth and the Sun. Transits are a bit like eclipses, but the whole Sun is not blocked out. Like eclipses, they happen only when the conditions are just right. The next transit of Venus will not be until 2117!

Venus and Mercury can be seen just before sunrise and just after sunset. At the right time, with clear skies, you can see them in the sky with other planets — or even the Moon.

Planets that are farther away from the Sun than Earth (Mars, Jupiter, Saturn, Uranus, and Neptune) are called **superior planets** and can appear anywhere in the sky along the ecliptic.

You can watch the planets all year as they swap and shuffle positions along the ecliptic. The planetary lineup is constantly changing!

Mars

Venus

Mercury

SPACE JOURNEYS

GREATEST ELONGATION. The point when Mercury or Venus is farthest from the Sun, as seen from Earth, very bright and easy to find.

OPPOSITION. The point when a superior planet is on the opposite side of the Earth from the Sun. Like the full Moon, it is brightest then, and up all night, from sunset to sunrise.

TRANSIT. The journey of Mercury or Venus across the face of the Sun.

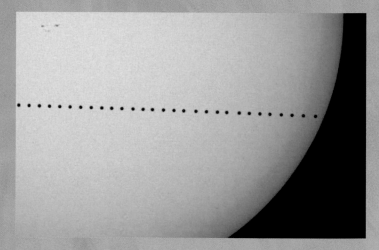

A **CONJUNCTION** (meaning "joining") is when two solar system objects travel close to each other in the sky. Although they may look as though they are in the same place, they are still very far from each other. It's simply that they are lined up in the ecliptic plane as seen from Earth.

SKY WANDERER

The word **planet** comes from *planetai*, a Greek word meaning "wanderer." In very ancient times, a planet was thought to be an object that wandered through the constellations. So Mercury, Venus, Mars, Jupiter, and Saturn were all considered planets, and so were the Sun and Moon!

Earth was not thought of as a planet at all — it was just Earth, the center of the universe and the place where we live.

Sun-Centered

When people figured out that Earth was not the center of the universe, the definition of the word *planet* changed to mean an object that orbits the Sun. The Sun and Moon were taken off the planet list, and Earth was added.

After the telescope was invented, astronomers discovered Uranus in 1781. Ceres was discovered in 1801, followed rapidly by Pallas, Juno, and Vesta. By 1807, astronomers had counted 11 planets!

Too Many Planets

Ceres, Pallas, Juno, and Vesta orbit the Sun between Mars and Jupiter. Their orbits cross each other, and as time went on, astronomers kept finding more objects in the same region.

By the 1850s, it was pretty clear that Ceres and its friends were not planets but a totally new type of solar system object, which they named **asteroids**. The number of planets slid from 45 to a more manageable seven.

Enter Neptune, Pluto, and Friends

An eighth planet, Neptune, was discovered in 1846. Then, in 1930, came the discovery of Pluto. It was the only new planet for a long time, until more objects were discovered out beyond Neptune's orbit starting in 1992.

The discovery of Eris ("air-iss") in 2005, an object almost as massive as Pluto, confirmed that Pluto was simply the brightest member of another new group of objects in the solar system, the **trans-Neptunian objects** (TNOs) .

Today, we define a planet as an object that is massive enough for its gravity to crush it into a sphere and that does not share its orbit with another object. Pluto, Eris, and Ceres are **dwarf planets**: objects that are spherical but share their orbits with other asteroids or TNOs. Two more TNOs, Haumea ("how-may-uh") and Makemake ("mocky-mocky"), are also dwarf planets, and astronomers expect many more to be discovered.

Earth is the center of the universe in this drawing, published in 1660, with the paths of the Sun, the Moon, five planets, and the zodiac circling it. By then many scientists, including Copernicus, Kepler, and Galileo, believed that Earth and the planets orbited the Sun instead, but the idea wasn't yet shared by everyone.

The first telescope was invented in 1608 by Hans Lippershey, a Dutch maker of eyeglasses, and news of it quickly spread to astronomers far and wide.

Roaming around the SOLAR SYSTEM

The solar system is not just a jumble of objects orbiting the Sun. Its organization can tell us about its formation and history. (The tilts of the orbits in this diagram are exaggerated.)

The **Oort Cloud** surrounds our entire solar system like a giant bubble made of icy planetesimals, left over from the formation of our Sun and planets.

Our Sun

The **Kuiper Belt** (rhymes with piper) is a disk that circles around our Sun on the ecliptic from beyond Neptune. Pluto lives in the Kuiper belt.

Scientists believe most comets are born in these distant regions.

The **Outer Solar System** is on a much larger scale than the Inner Solar system and contains the Jovian gas giant planets, Jupiter, Saturn, Uranus, and Neptune.

Saturn

SUN

Jupiter

Uranus

Neptune

The **Inner Solar System** lies within the Asteroid Belt and consists of the Sun, the rocky terrestrial planets, and their moons.

Mercury

Mars

Venus

Earth

SUN

Asteroid belt

Jupiter

Mercury

Earth

Venus

Mars

Asteroid Belt

Saturn

Uranus

Neptune

Kuiper Belt

Terrestrial Planets

Mercury, Venus, Earth, and Mars are called **terrestrial** planets. Most have atmospheres. All are pretty close to the Sun (1.5 AU or less) and have few or no moons.

Terrestrial means "Earth-like." All the terrestrial planets are made of rock and metal and are the size of Earth or smaller.

Space Talk

AU (Astronomical Unit): A unit used for measuring distances. One AU is equal to the distance of Earth from the Sun (93 million miles [150 million km])

Jovian Planets

Jupiter, Saturn, Uranus, and Neptune are the **jovian** planets. They are farther out in the solar system (5 to 30 AU from the Sun).

Jovian means "Jupiter-like." They are four to eleven times the size of Earth and have much more mass than the terrestrial planets. Unlike the terrestrial planets, jovian planets don't have a hard surface that a spacecraft can land on. They are made of gases.

Jovian planets all have rings and lots of moons.

Small Solar System Bodies (SSSBs)

There are lots of smaller objects in the solar system. These chunks of ice and rock orbit the Sun in three main regions.

The **asteroid belt** (1.5 to 5 AU from the Sun) lies between the orbits of Mars and Jupiter. It contains asteroids, large chunks of rock that orbit the Sun.

ASTEROID "BENNU"

Most asteroids are "rubble piles" — rock collections loosely held together by gravity.

The Kuiper belt (30 to 50 AU from the Sun) contains objects made of ice and rock.

The **Oort cloud** (1,000 to 100,000 AU from the Sun) is a spherical shell of icy objects surrounding the solar system. It extends nearly a quarter of the way to our nearest neighbor star, Proxima Centauri.

Meteoroids are chunks of rock and ice found outside the asteroid belt, Kuiper belt, or Oort cloud. Sometimes these meteoroids are pulled in by a planet or moon's gravity.

Meteors, "shooting stars" that we sometimes see in our night-time sky, are meteoroids entering our atmosphere. Any bit of the meteor that survives to crash-land on a planet is called a **meteorite**.

MAKE A SCALE MODEL ✷ ✷

The distances between planets in our solar system are huge compared to the sizes of the planets. It would be impossible to draw the entire solar system to scale on one piece of paper.

To see this for yourself, you can make a *scale model* of the solar system — an exact copy, shrunk down in size.

1. Start with a Sun that is 6 inches (15 cm) across — about the size of a large grapefruit. This is about one 10-billionth of its true size. Use the table to figure out how big each planet is and how far it should be from the model Sun.

2. Find objects around your house to represent each planet. (A mustard seed would work for Earth.)

3. You will need to head outside to have enough space for your model solar system. You can have your friends hold the model planets and spread themselves out in a field or on the sidewalk.

Your solar system will reach only as far as Jupiter if your model is on a football field. Check with a web-based map program to see where the rest of the planets would end up in your town.

With a scale model this size, the nearest star (Proxima Centauri) will be another grapefruit 1,900 miles (3,000 km) away. Space is a very empty place!

OBJECT	DIAMETER			USE TO COMPARE	DISTANCE FROM THE SUN	
	mm	inches	inches		meters	feet
SUN	150.0	6.00	6	grapefruit		
MERCURY	0.5	0.02	1/32	grain of salt	6.2	20
VENUS	1.3	0.05	1/16	mustard seed	11.6	38
EARTH	1.4	0.05	1/16	mustard seed	16.0	53
MARS	0.7	0.03	1/32	poppyseed	24.4	80
JUPITER	15.3	0.60	5/8	small red grape	83.4	274
SATURN	12.9	0.51	1/2	blueberry	152.9	502
URANUS	5.5	0.22	1/4	peppercorn	307.6	1009
NEPTUNE	5.3	0.21	1/4	peppercorn	481.9	1581
PLUTO	0.3	0.01	1/64	grain of salt	632.8	2076

It's easier to keep track of your tiny planets if you tape them onto separate index cards. For Mercury and Pluto, just make a dot on the card with a fine-tip pen!

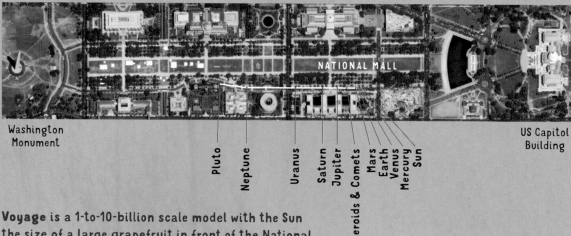

Washington Monument

US Capitol Building

Pluto Neptune Uranus Saturn Jupiter Asteroids & Comets Mars Earth Venus Mercury Sun

Voyage is a 1-to-10-billion scale model with the Sun the size of a large grapefruit in front of the National Air and Space Museum in Washington, DC. The planets are placed along the National Mall. There are Voyage models in Kansas City, Missouri, and in Corpus Christi and Houston, Texas.

MERCURY

AVERAGE DISTANCE FROM THE SUN:
0.39 astronomical units (AU)

DIAMETER (COMPARED TO EARTH): 0.38

MADE OF: rock and metal

ATMOSPHERE: none

MOONS: none

RINGS: none

TEMPERATURE: -280 to
800°F (-170 to 430°C)

1 DAY = 176 EARTH DAYS

1 YEAR = 88 EARTH DAYS

WHAT IS IT LIKE ON MERCURY?
No air, lots of craters. Freezing
at night and super-hot during
the very, very long day.

HAVE ANY SPACECRAFT VISITED IT?
NASA's Mariner 10 and MESSENGER;
ESA-JAXA's BepiColombo

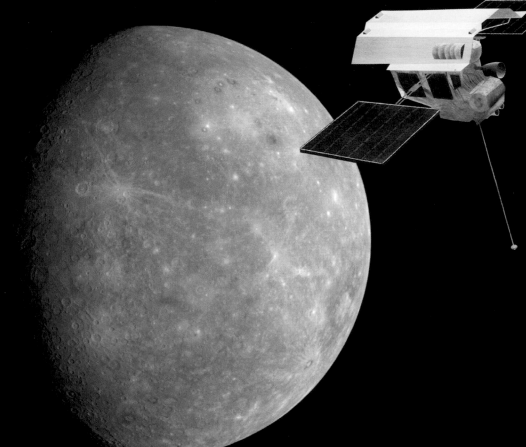

MESSENGER was
launched from Earth
on August 3, 2004,
began its orbit of
Mercury on March 17,
2011, and completed
its mission on April
30, 2015.

That's no moon... Mercury
looks like Earth's Moon,
but it's a little larger
and much more massive.

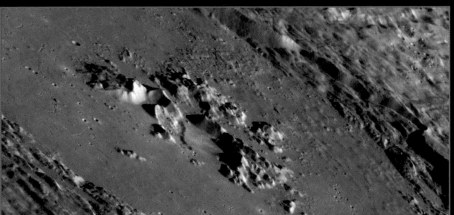

MERCURY IS COVERED WITH CRATERS. This close-up of Abedin crater shows
the debris from a massive meteoroid strike. The rock melted during the collision,
and the splash solidified in a jumbled mass at the crater's center.

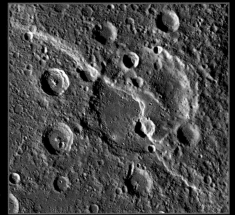

SCARPS are wrinkles in Mercury's
crust caused by its interior cooling and
shrinking.

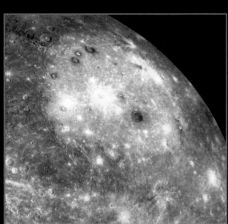

The colors in this photo show
different types of rock. Tan indicates
volcanic rock.

VENUS

AVERAGE DISTANCE FROM THE SUN: 0.72 AU

DIAMETER (COMPARED TO EARTH): 0.95

MADE OF: rock and metal

ATMOSPHERE: carbon dioxide and nitrogen

MOONS: none

RINGS: none

TEMPERATURE: 860°F (460°C) all day

1 DAY = 117 EARTH DAYS

1 YEAR = 225 EARTH DAYS

WHAT IS IT LIKE ON VENUS? With dense clouds, sulfuric acid rain, and a thick atmosphere that can crush you in minutes, Venus is not a nice place to visit. Also, there's no water.

HAVE ANY SPACECRAFT VISITED IT? Venera 4–16; Mariner 2, 5, and 10; Pioneer Venus 1 and 2; the Magellan orbiter; and ESA's Venus Express

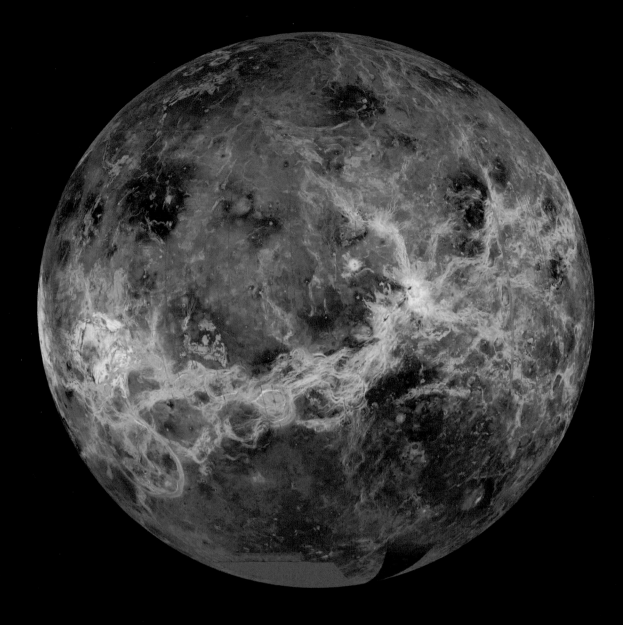

What is Venus hiding underneath these thick clouds?

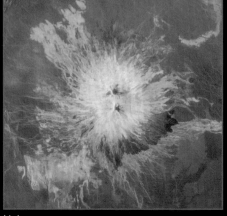

Volcanoes ...

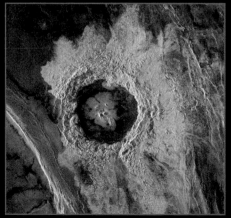

craters ...

... and a rocky surface.

Most features on Venus are named for goddesses or women, like this crater named for poet Emily Dickinson.

EARTH

AVERAGE DISTANCE FROM THE SUN: 1 AU, or 93,000,000 miles (150,000,000 km)

DIAMETER: 7,900 miles (12,700 km)

MADE OF: rock and metal

ATMOSPHERE: nitrogen, oxygen, and carbon dioxide

MOONS: 1

RINGS: none

TEMPERATURE: -126 to 136°F (-88 to 58°C)

1 DAY = 24 EARTH HOURS

1 YEAR = 365.25 EARTH DAYS

WHAT IS IT LIKE ON EARTH? Comfortable temperatures and a breathable atmosphere make this the solar system's garden planet. It has liquid water on its surface and life everywhere.

HAVE ANY SPACECRAFT VISITED IT? Lots of spacecraft orbit the Earth, studying its weather and taking pictures of the surface. There's even an orbiter with people living in it!

Earth is at exactly the right temperature to have a water cycle.

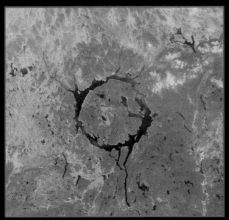

Most craters on Earth have been erased by weather and plate tectonic action.

Of all the moons in the solar system, Earth's Moon is the largest in relation to its planet.

After 4 billion years of evolution, Earth's life is pretty advanced!

MARS

AVERAGE DISTANCE FROM THE SUN: 1.5 AU

DIAMETER (COMPARED TO EARTH): 0.53

MADE OF: mostly rock

ATMOSPHERE: carbon dioxide, nitrogen, and argon

MOONS: 2, Phobos and Deimos

RINGS: none

TEMPERATURE: -125 to 135°F (-90 to 60°C)

1 DAY = 24.6 EARTH HOURS

1 YEAR = 687 EARTH DAYS

WHAT IS IT LIKE ON MARS? Colder than Earth, with a very thin atmosphere and low gravity. The soil and rocks are reddish, thanks to iron in the soil.

HAVE ANY SPACECRAFT VISITED IT? Many, including rovers that drive around the planet.

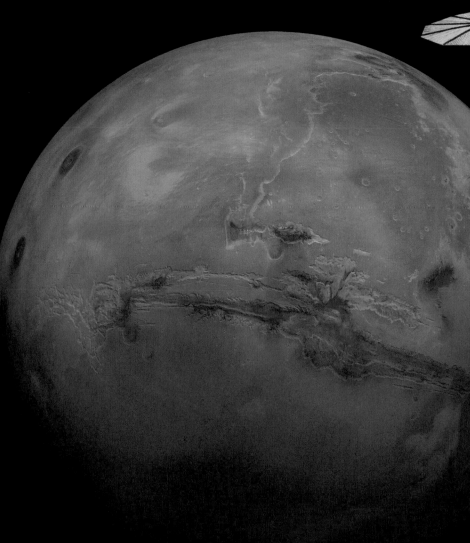

INSIGHT launched from Earth on May 5, 2018, landed on Mars on November 26, 2018, and continues to investigate what lies beneath the Martian surface.

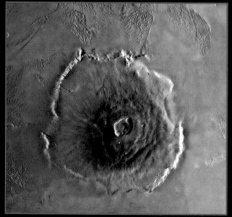

Olympus Mons is the tallest volcano in the solar system.

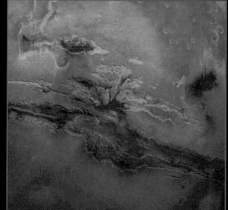

Valles Marineris is the largest canyon in the solar system.

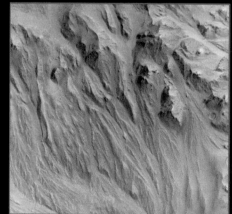

Most of Mars's water is frozen in its polar ice caps, but this photo shows evidence that there may once have been liquid water on the surface.

Mars's two moons, Deimos (left) and Phobos (right), are probably asteroids that wandered too close to Mars and were pulled into orbit by its gravity.

SPACE ODYSSEYS

TELESCOPES IN ORBIT AROUND EARTH see objects in space more clearly than those on the ground because Earth's atmosphere is not in their way. (Orbit means the path that an object takes around another object.)

SPACE PROBES, which journey farther out into space, can see even better. Most space probes are designed for a one-way trip: they collect information and transmit it back to Earth.

ORBITERS circle a planet, moon, or asteroid and take lots of pictures.

LANDERS touch down on the surface of a planet, moon, or asteroid to take really close-up pictures and examine its rocks, soil, and atmosphere.

ROVERS are landers that can drive around on a planet's surface and explore the terrain.

SAMPLE-RETURN MISSIONS are spacecraft that visit a planet or other solar system body, collect a sample of rock, and return the rock to Earth for us to study.

INTERSTELLAR SPACECRAFT are probes designed to leave the solar system and explore the space between star systems.

CURIOSITY launched from Earth on November 26, 2011, landed on Mars on August 6, 2012, and continues to roam the planet.

MARS SELFIE: The Mars Curiosity rover took this "selfie" on January 23, 2018. It was compiled from a few dozen photos taken with a camera in the "hand" of its robot arm.

JUPITER

AVERAGE DISTANCE FROM THE SUN: 5.2 AU

DIAMETER (COMPARED TO EARTH): 11

MADE OF: hydrogen and helium

ATMOSPHERE: hydrogen and helium

MOONS: 79 that scientists have counted so far

RINGS: very thin

WHAT IS IT LIKE ON JUPITER? Jupiter doesn't have a solid surface to stand on. It is made of gas and covered with thick clouds.

TEMPERATURE = -230°F (-150°C)

1 DAY = 10 EARTH HOURS

1 YEAR = 11.8 EARTH YEARS

HAVE ANY SPACECRAFT VISITED IT? Pioneer 10 and 11, Voyager 1 and 2, and Juno. Cassini and New Horizons flew by on their way to Saturn and Pluto.

JUNO was launched from Earth on August 5, 2011, entered Jupiter's orbit on July 4, 2016, and completes its mission in July 2021.

The Great Red Spot is a giant storm that's been around for at least 350 years.

The stripes on Jupiter are clouds of ammonia (white) and ammonium hydrosulfide (red).

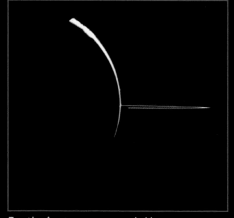

Jupiter's rings are much thinner than Saturn's.

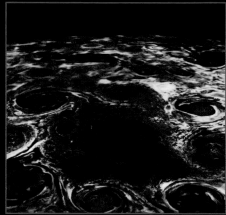

Jupiter's north pole is a cyclone, surrounded by eight more cyclones. (Its south pole looks similar, but with

Galilean Moons

Jupiter's four largest moons, first spotted by Galileo in 1609, are tidally locked to Jupiter — they always show Jupiter their same side. The closest to Jupiter is Io, then Europa, Ganymede, and Callisto. You can remember their order in the Jupiter system with the mnemonic "I Eat Green Cheese."

The other moons of Jupiter are small, with elliptical (oval), tilted orbits. They are most likely captured asteroids and comets. Jupiter has shielded the inner solar system from a lot of giant meteor impacts!

Io   Europa   Ganymede   Callisto

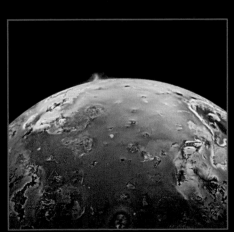

Io's volcanoes are so strong that they spew material into Io's orbit.

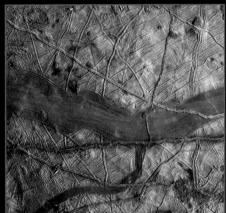

The moon Europa has a rocky core, an icy surface, and possibly an underground ocean. The red stripes are cracks where salty underground water may have leaked onto the surface.

HELLO, UNIVERSE!

Probably the most successful solar system explorers were Voyager 1 and 2. Launched in 1977, the Voyagers made a grand tour of Jupiter, Saturn, Uranus, and Neptune in only 12 years. They changed everything we know about these planets and their moons.

Both Voyagers have now left the solar system, becoming interstellar probes. Both carry what is called the Golden Record — a gold-plated copper disk containing pictures and sound files that tell the story of human civilization. It's meant to be a message for any extraterrestrials the probes might run into. The record's cover has instructions for non-human scientists to build a record player.

The Golden Record carries greetings in 55 different languages and music from around the globe, as well as images of our world and human culture. The disk also contains Earth noises, such as thunder and a volcanic eruption; animal sounds, from cricket chants to bird songs to elephant trumpetings; and human sounds, such as laughter, a heartbeat, singing, and speech.

SATURN

AVERAGE DISTANCE FROM THE SUN: 9.5 AU

DIAMETER (COMPARED TO EARTH): 9.1

MADE OF: hydrogen and helium

ATMOSPHERE: hydrogen and helium

MOONS: 82

RINGS: amazing

DAYTIME TEMPERATURE: -290°F (-180°C)

1 DAY = 10.7 EARTH HOURS

1 YEAR = 29 EARTH YEARS

WHAT IS IT LIKE ON SATURN? Like Jupiter, it is made of gas and covered with thick clouds. Winds on Saturn can reach speeds of 1,100 mph (1,800 kph).

HAVE ANY SPACECRAFT VISITED IT? Pioneer 11, Voyager 1 and 2, and Cassini-Huygens. Dragonfly will arrive in the 2030s.

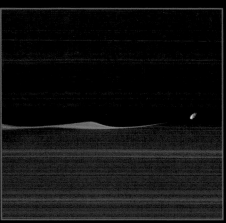

"Shepherd moons" like Daphnis help keep the gaps in Saturn's rings clear.

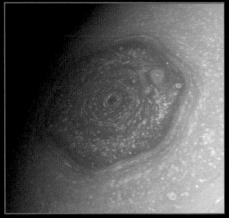

There's a hexagon-shaped storm at Saturn's north pole.

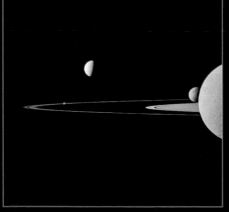

The Saturn moon system is crowded! This picture taken by Cassini shows five moons in one photo.

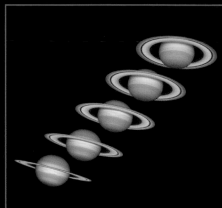

RING TILT: Because Earth and Saturn tilt relative to each other, we see Saturn's rings at different angles over time.

DRAGONFLY

Titan

Titan, the second largest moon in the Solar System, is bigger than Mercury. It has a thick nitrogen-methane atmosphere that hides its surface. The Cassini space probe dropped the Huygens lander on Titan and found methane and ethane (natural gas) lakes, clouds, and rain. Titan has a rocky core, a surface made of rock-hard ice, and probably a salty underground ocean made of water and ammonia.

The Dragonfly lander drone will travel to Titan in the 2030s.

TITAN: The best close-up we have of Titan's surface is from ESA's Huygens lander. Its resolution is low, but it shows ice pebbles under an orange sky.

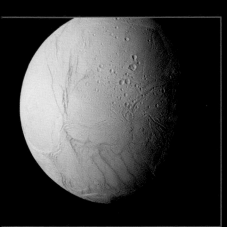

Saturn's moon Enceladus looks a lot like Jupiter's moon Europa. Scientists think there may be an underground ocean there, too!

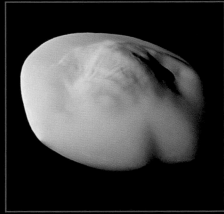

Moons that orbit in Saturn's rings get covered with dust and ice, which makes them look like ravioli.

DESIGN YOUR OWN SOLAR SYSTEM ✦✦

Astronomers have discovered thousands of planets orbiting other stars. These exoplanets can be much different from Earth and its neighbors.

WHAT YOU NEED:

- Astronomy Notebook
- Colored pencils, pens, markers, or crayons

1. Start with the star that will be your sun. Decide what color it will be, and draw it.

2. How many planets will you have, and how will they be different from one another? Rocky planets are usually close to the star and gaseous planets are farther out. Add them to your drawing.

3. Think about where the "habitable zone" is. This is the region in a solar system where water is liquid and there might be life! What kinds of plants, animals, and maybe even aliens live in your solar system? Draw them on another page.

4. What else is in your solar system? Gaseous planets that have wandered close to the sun? Comets and asteroids whizzing around? Alien space stations?

5. Make up names for the star and all of its planets and moons.

→ **What would it be like to be an alien?**

→ **What would your aliens think of us Earthlings?**

Let your imagination go wild!

URANUS

AVERAGE DISTANCE FROM THE SUN: 19 AU

DIAMETER (COMPARED TO EARTH): 4.0

MADE OF: hydrogen and helium

ATMOSPHERE: hydrogen, helium, and methane

MOONS: 27

RINGS: yes – second-largest in the solar system

DAYTIME TEMPERATURE: -360°F (-215°C)

1 DAY = 17.2 EARTH HOURS

1 YEAR = 84 EARTH YEARS

WHAT IS IT LIKE ON URANUS?
The surface is similar to an icy ocean. Uranus is lying on its side compared to the other planets, and that causes extreme seasons.

HAVE ANY SPACECRAFT VISITED IT? Voyager 2

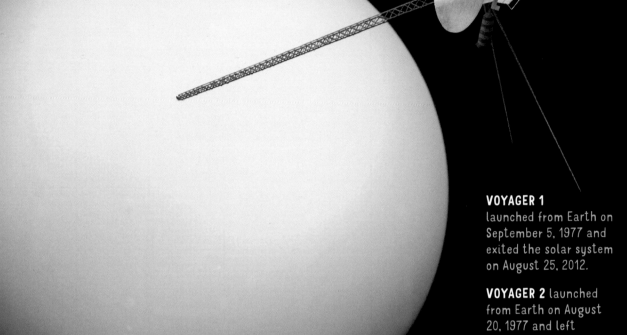

VOYAGER 1 launched from Earth on September 5, 1977 and exited the solar system on August 25, 2012.

VOYAGER 2 launched from Earth on August 20, 1977 and left the solar system on November 5, 2018. Both missions are ongoing.

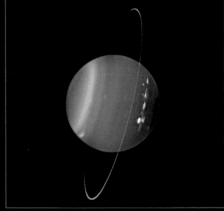

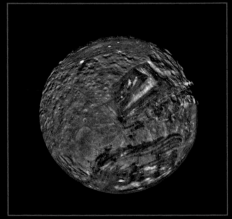

When Voyager 2 visited Uranus during its Northern Hemisphere summer, it looked like a smooth, turquoise sphere. Photos taken by the Hubble telescope (left, 1998) and the Keck telescope (right, 2004) near Uranus's fall season show stripes and storms, as well as Uranus's rings.

Titania is Uranus's largest moon.

Uranus's moon Miranda looks like a bunch of giant boulders smashed together.

NEPTUNE

AVERAGE DISTANCE FROM THE SUN: 30 AU

DIAMETER (COMPARED TO EARTH)**:** 3.9

MADE OF: Hydrogen and helium

ATMOSPHERE: Hydrogen, helium, and methane

MOONS: 14

RINGS: yes

DAYTIME TEMPERATURE: -360°F (-215°C)

1 DAY = 16.1 EARTH HOURS

1 YEAR = 165 EARTH YEARS

WHAT IS IT LIKE ON NEPTUNE? Like Uranus, but darker. It is dark, cold, and windy, and the surface is a slushy soup of methane (natural gas).

HAVE ANY SPACECRAFT VISITED IT? Voyager 2

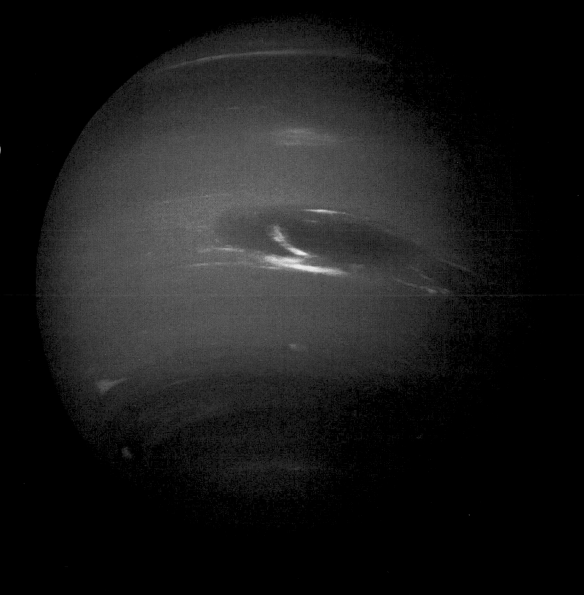

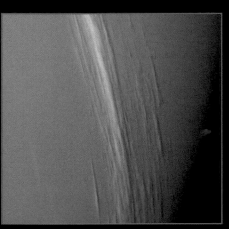

Like the other jovian planets, Neptune has cloudy stripes.

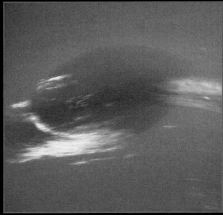

THE GREAT DARK SPOT is an Earth-sized storm that Voyager 2 observed in 1989. It has disappeared since then.

NEPTUNE'S MOON TRITON is the seventh largest moon in the solar system. It resembles Pluto and orbits backwards. It may be a TNO (see page 65) that strayed too close to Neptune.

When Triton was captured, it probably disrupted the whole moon system, leaving only a few small moons, like Proteus (left) and Larissa (right).

PLUTO

AVERAGE DISTANCE FROM THE SUN: 39.5 AU

DIAMETER (COMPARED TO EARTH)): 0.19

MADE OF: Ice and rock

ATMOSPHERE: Nitrogen, methane, and carbon monoxide

MOONS: 5

RINGS: None

DAYTIME TEMPERATURE: -375° (-226°)

1 DAY = 153 EARTH HOURS

1 YEAR = 248 EARTH YEARS

WHAT IS IT LIKE ON PLUTO? Very cold and dark. From here, the Sun looks like a very bright star. Pluto and its largest moon, Charon, are tidally locked together. Charon shows only one face to Pluto (like Earth's Moon does), and Pluto shows only one face to Charon.

HAVE ANY SPACECRAFT VISITED IT? New Horizons

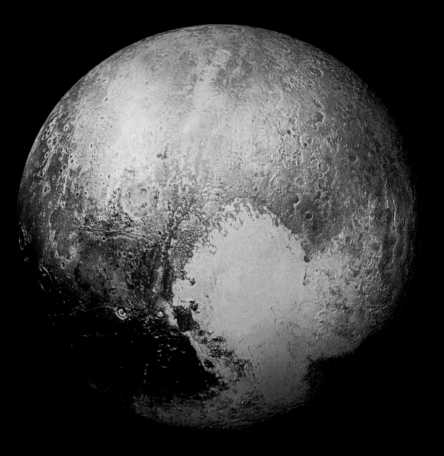

NEW HORIZONS launched from Earth on January 19, 2006, and executed its flyby of Pluto in 2015 and 2016. Its mission is ongoing, as it travels through the Kuiper belt and beyond.

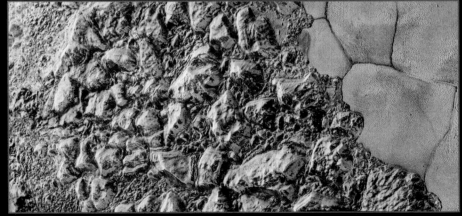

This close-up was taken by the New Horizons spacecraft. It shows craters, mountains, and glaciers on Pluto's surface.

PLUTO has an atmosphere when its elliptical orbit brings it slightly closer to the Sun. Farther from the Sun, the air and clouds freeze and fall to the ground.

Pluto's largest moon, Charon, would be a dwarf planet if it hadn't been captured by Pluto.

Meet the Other Dwarf Planets

Some objects in the asteroid belt and Kuiper belt are so large that gravity has crushed them into spheres. These are the dwarf planets.

	AVERAGE DISTANCE FROM THE SUN (IN AU*)	YEAR (IN EARTH YEARS)	DAY (IN EARTH DAYS)	MADE OF
CERES	2.8	4.6	9.0	rock
HAUMEA	43.0	281.9	3.9	ice
MAKEMAKE	45.3	305.3	22.5	ice
ERIS	68.0	561.4	25.9	ice

AU =astronomical units. 1 AU is the distance from the Sun to Earth.

 Only the photo of Ceres is real. Spacecraft have not visited Haumea, Makemake, or Eris, so these images are the best guess as to what they look like.

BE THE SOLAR SYSTEM ★ ★ ★

If you have a big group of people and an open area (schoolyard or athletic field) you can act out how the Sun, the planets, and their moons all move together as if in a cosmic game or dance.

1. Give each player a name tag or slip of paper with the name of a solar system object. Start with our Moon and all of the planets. If you have enough people you can include Mars's two moons (Phobos and Deimos) and the most famous moons of Jupiter (Europa, Ganymede, Io, and Callisto), Saturn (Titan and Dione), and Neptune (Triton).

2. In a central spot, place a marker such as a hat or a backpack to serve as the Sun.

3. Line up all planets in order, from Mercury to Neptune, moving out from the Sun. All players who are portraying moons should join their planet.

4. When you are ready, signal all players playing moons to begin circling their planet.

5. Signal all players playing planets to proceed in a slow circle counterclockwise around the sun, keeping the same distance, with the moons continuing to circle around their planets.

6. See how long you can go before you get tangled up or collapse in laughter. And remember, our solar system has been doing this for billions of years!

HOW THE SOLAR SYSTEM FORMED

When the Sun began to form, it was surrounded by a spinning disk of gas and dust. As the disk spun, small bits of matter smashed together and grew bigger, turning into solid bodies called planetesimals and then larger masses called **protoplanets** ("early planets"). These objects were so massive that gravity shaped them into spheres.

Cool & Hot

The Frost Line is an imaginary circle around the Sun. Inside the circle, closer to the sun, it was warm enough that water and other **elements** evaporated, leaving behind only rocks and minerals. Protoplanets forming inside the Frost Line became the rocky terrestrial planets.

Outside the Frost Line, it was so cold that water and other elements froze into ice.

Protoplanets here formed into the jovian planets, growing more massive as their gravity pulled in more dust and gases. They took shape like small solar systems, with a giant planet in the center surrounded by a disk that formed the larger moons.

Things Settle Down

All this time the Sun was still forming (page 52). As it got hotter, its heat and the solar wind (page 22) cleared gases and smaller ice particles from the solar system.

The terrestrial planets became solid. The jovian planets moved to where they are now. Most of the icy planetesimals moved into the Kuiper belt and the Oort cloud. Some were captured by the jovian planets and became moons.

The solar system's formation was complete!

How Do We Know?

To figure out how the solar system came to be, scientists look at how it is organized (its structure) and what it is made of (its composition). Here is what we know.

★ The planets' orbits are nearly circular and all near the same plane.

★ The planets all orbit in the same direction. Most of them spin that way, too, as do most of their moons — and the Sun itself.

★ Planets closer to the Sun are small and rocky; planets farther from the Sun are very large and made of compounds containing hydrogen.

★ All planets are differentiated, with dense materials in their core and less dense material at the surface.

★ The solar system also contains small, rocky bodies close to the Sun (the asteroid belt) and small, icy bodies farther from the Sun (the Kuiper belt and Oort cloud).

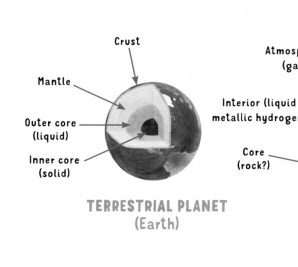

Crust
Mantle
Outer core (liquid)
Inner core (solid)

TERRESTRIAL PLANET
(Earth)

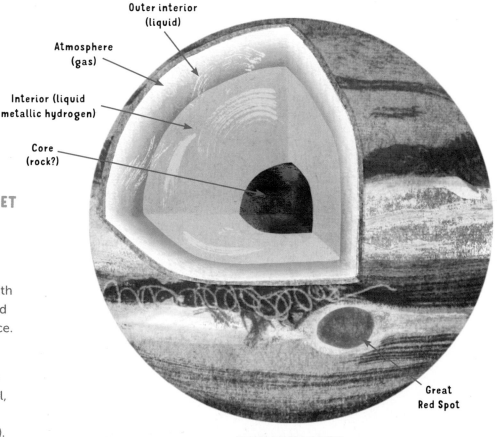

Outer interior (liquid)
Atmosphere (gas)
Interior (liquid metallic hydrogen)
Core (rock?)
Great Red Spot

JOVIAN PLANET
(Jupiter)

Uranus

BIRTH OF OUR SOLAR SYSTEM

CRASH!

4.5 BILLION YEARS AGO

The Sun forms in the center of a spinning disk of gas and dust.

As the disk spins, dust grains stick together due to gravity and form pebbles, then rocks, then planetesimals.

Planetesimals smash and melt together and form spherical planets, clearing out gaps in the disk.

BANG!

ROCKY PLANETESIMALS

FROST LINE

Outside the frost line, the protoplanets become the larger jovian planets, made of hydrogen and other gases.

ICY PLANETESIMALS

Inside this imaginary circle around the sun, the protoplanets become the smaller terrestrial planets, made only of rocks and minerals. All other substances such as water have evaporated.

TADA!

The new, hot sun blows away all remaining gas and dust.

Leftover rocky planetesimals form the ASTEROID BELT

Leftover icy planetesimals form the KUIPER BELT

GREAT COMET

Comets are large chunks of ice and rock (about the size of a mountain) whose long, skinny orbits bring them close to the Sun and then out to the far edge of the solar system. They come from the Kuiper belt or the Oort cloud, usually when the gravity of a jovian planet knocks them out of orbit.

Comets used to be thought of as bad omens. Most ancient civilizations made predictions based on the motions of objects in the sky. Any change in the stars and planets could mean trouble on Earth!

Nowadays, it's exciting to see a comet. We know what they are, and when one visits the inner solar system, it's a special occasion for sky gazers.

Parts of a Comet

When a comet comes close to the Sun — within about 5 AU — the ice on the outside turns to vapor. The remaining chunk of rock and ice becomes the nucleus of the comet, and the **vaporized** gases form the atmosphere, or **coma**.

When a comet gets to within about 1 AU of the Sun, it forms two tails. These tails are longest and brightest when the comet is closest to the Sun.

The ion tail (sometimes called a gas tail or plasma tail) is made of ionized gas from the comet's atmosphere. It is shaped by magnetic forces in the solar wind and always faces away from the Sun.

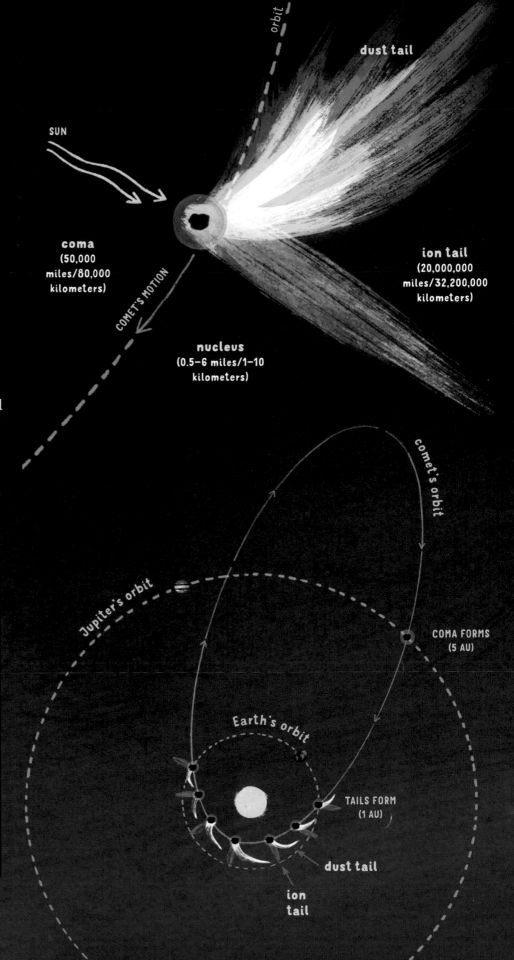

orbit

dust tail

SUN

coma (50,000 miles/80,000 kilometers)

COMET'S MOTION

ion tail (20,000,000 miles/32,200,000 kilometers)

nucleus (0.5–6 miles/1–10 kilometers)

comet's orbit

Jupiter's orbit

COMA FORMS (5 AU)

Earth's orbit

TAILS FORM (1 AU)

dust tail

ion tail

What's at the heart of a comet? The nucleus of Comet 67P/ Churyumov-Gerasimenko looks a lot like an asteroid. The streaks are jets of gas and dust erupting as the comet melts.

The dust tail is made of dust particles that are freed from the comet's nucleus as the ice melts. These particles are so light that they are pushed outward by sunlight! The dust tail sticks out away from the Sun but also curves along the comet's path.

Some of the dust is too heavy to be pushed by light and stays in the comet's orbit. If Earth crosses the comet's orbit, these dust particles will become meteors (see page 8).

Long-Distance Travelers

Comets from the Oort cloud have random orbits and come from all directions. They usually take thousands of years to complete one orbit.

Comets from the Kuiper belt orbit in the plane of the solar system in the same direction as the planets, with journeys of less than 200 years.

More than 3,500 comets have been discovered — so far. Most comets stay in the far reaches of the solar system, never coming close enough to the Sun to form tails and frighten superstitious humans.

A great comet is one that is visible with the naked eye. Whether a comet is great depends on its size, composition, and distance from the Sun and Earth. It's very hard for astronomers to predict if a comet will be great. On average, a great comet passes by about every 10 years.

How to Observe Comets

If a comet visits our part of the solar system, we usually hear about it in the news. From night to night, it will change position compared to the stars around it. Check the web for a map of where it is among the constellations and its rising and setting times. A comet's tail may make it look like it's shooting across the sky, but comets move slowly around the Sun, just like planets do.

Go outside and observe the comet whenever it's clear, and draw pictures in your Astronomy Notebook. You can watch as the tail grows, shrinks, and changes shape.

 The word *comet* comes from the Greek *aster kométēs*, which means "hairy star."

Comet Hale-Bopp, the great comet of 1997. The blue tail is the ion tail, and the yellow tail is the dust tail.

The last great comet was C/2006 P1 (McNaught), which was visible in 2007 but only in the Southern Hemisphere. This image shows the beautiful dust tail fanning out, with streamers from the dust's interaction with the solar wind. (The bright object to the right is the Moon.)

OTHER SUNS AND THEIR SOLAR SYSTEMS

We now think that most stars form solar systems. Astronomers call a planet that orbits a star other than our Sun an **exoplanet**.

It's very hard to find exoplanets. Stars are so big and bright that planets are nearly invisible next to them. Nevertheless, astronomers have found exoplanets orbiting thousands of stars. Here are four main methods for finding them.

★ **THE TRANSIT METHOD** looks for stars that get dimmer when a planet transits (passes in front of) them.

★ **THE IMAGING METHOD** looks at stars in different colors of light and uses computer image processing to block the light of the star (see below).

★ **THE DOPPLER METHOD** looks for changes in a star's velocity (speed) as it is pushed and pulled by an orbiting planet.

★ **THE MICROLENSING METHOD** looks for distant stars that get *brighter* when another star and planet pass in front of them. The brightening happens because the gravity of the nearby star and planet acts as a lens to magnify the light from the distant star.

Most of the solar systems we have found are very unlike our own. We have found super-Earths, mini-Neptunes, and jovian planets close to their suns. Our own solar system would be hard to find using these methods, so we don't yet know if our solar system is typical or an oddball.

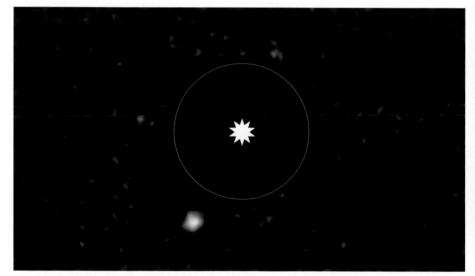

The bright light of the **star** HD 95086 was subtracted out of this image by a computer, leaving behind an image of its **planet**, called HD 95086 b. The white star marks the position of HD 95086, and the blue circle around it has a radius of 30 AU — equivalent to the distance from our Sun to Neptune.

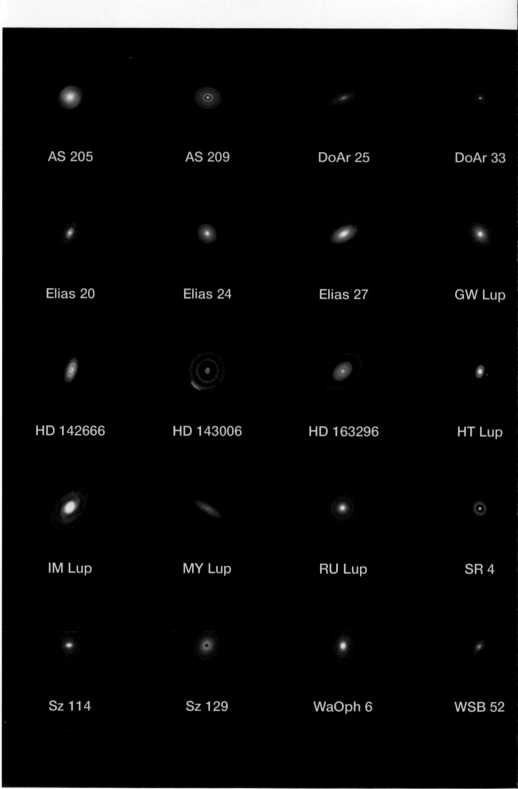

Protoplanetary disks around some stars, observed by a radio telescope. Astronomers think the dark spaces may be gaps in the disks caused by planet formation.

The Planets

Like the Moon, Venus goes through phases as it revolves around the Sun. Through binoculars at certain times you can clearly see its crescent phase. Mercury too has phases, but it appears too small in our sky for us to see them.

Superior planets don't go through phases. The best time to see them is at opposition (see page 64). Mars looks like an orange star through binoculars. But if you compare it to a nearby star, it looks like a disk, not a point of light.

A composite image of pictures taken of Venus for a whole phase cycle around the Sun, from July 2010 to January 2012.

Through binoculars or a small telescope the moons of Jupiter look like dots, all lying in a straight line with their parent planet. They change position from day to day.

Comet Hale-Bopp, the great comet of 1997, was visible to the naked eye for 18 months. Its tail stretched over a quarter of the sky.

Moons of Jupiter

Jupiter also looks like a disk, but one that is a bit larger than Mars. With 50 mm binoculars, you can easily see the four largest moons of Jupiter: Io, Europa, Ganymede, and Callisto. They're visible even with smaller binoculars when they're far enough away from Jupiter in their orbits.

Draw your view of Jupiter over a few days to see the moons' motion. Apps and websites can tell you which moon is which.

Golden Saturn

Saturn looks like a yellowish oval through binoculars with a magnification of 10 or more. With less magnification, it's pretty circular. You can't really see the rings clearly without a telescope.

Planet Party

If two planets are in conjunction, you can see them together in your binoculars' field of view.

Great Comets

Some comets can be seen only through binoculars or a telescope. In binoculars, they look like faint fuzz balls. You can see great comets with your naked eye, but binoculars will show all kinds of interesting details in the coma and tail.

5

STARS and CONSTELLATIONS

On a clear, dark night, it's possible to see 4,500 stars. Most people can't see that many, of course! The clouds, Moon, and bright lights blot out the faintest stars. In the average suburb, you can see about 450 stars due to light pollution. In most big cities, you can see only about 35 stars — but you can learn to know them all!

STAR LIGHT, STAR BRIGHT

Stars make their own light by fusion, just the way the Sun does.

Why are some stars bright and some faint? How bright a star is in the sky depends on two things: the star's true brightness and its distance from us. A star that is faint and close can look brighter than a star that is bright and far away.

The Greek astronomer Hipparchus (c. 190–120 BCE) divided stars into six groups according to their brightness. Modern astronomers still use this system.

First-magnitude stars are the brightest stars, visible just after sunset. Sixth-magnitude stars are the faintest, visible only on a clear night with no Moon, away from city lights. The faintest stars we can see are magnitude 6.5.

✸	0TH MAGNITUDE	**Capella** in Auriga
✹	1ST MAGNITUDE	**Pollux** in Gemini
✶	2ND MAGNITUDE	**Polaris** (the North Star)
✦	3RD MAGNITUDE	**Gomeisa** in Canis Major
◆	4TH MAGNITUDE	**Epsilon Aquilae**

The 8 Brightest Stars in Our Sky

STAR	CONSTELLATION	HEMISPHERE
The Sun	---	---
Sirius	Canis Major (the Great Dog)	Both
Canopus	Carina (the Keel)	Southern
Rigil Kentaurus	Centaurus (the Centaur)	Southern
Arcturus	Boötes (the Herdsman)	Both
Vega	Lyra (the Lyre)	Northern
Capella	Auriga (the Charioteer)	Northern
Rigel	Orion (the Hunter)	Both

STARS OF ORION

Betelgeuse ⓪
Bellatrix ①
Alnilam ①
Mintaka ①
Alnitak ①
Great Nebula
Saiph ①
Rigel ⓪

Curious about the different colors of stars? See page 117.

HOW STARS MOVE DURING THE NIGHT

Imagine a crystal sphere surrounding Earth, with all of the constellations painted on it. This is how ancient people saw the sky, and it's still a good model for visualizing the stars' motions. As our Earth rotates from west to east, the constellations move from east to west.

The celestial poles are the points around which the stars seem to be spinning. Stars that are close to the celestial poles make small circles as the Earth rotates. Stars near the celestial equator make large circles.

The North Star

Polaris, the star at the end of the handle of the Little Dipper, is 0.5 degree from the North Celestial Pole. It barely moves at all during the night. We call it the North Star because no matter how the stars move, Polaris is always in the north.

Although the stars seem to move from east to west, like the Sun, it's really the Earth that's spinning west to east.

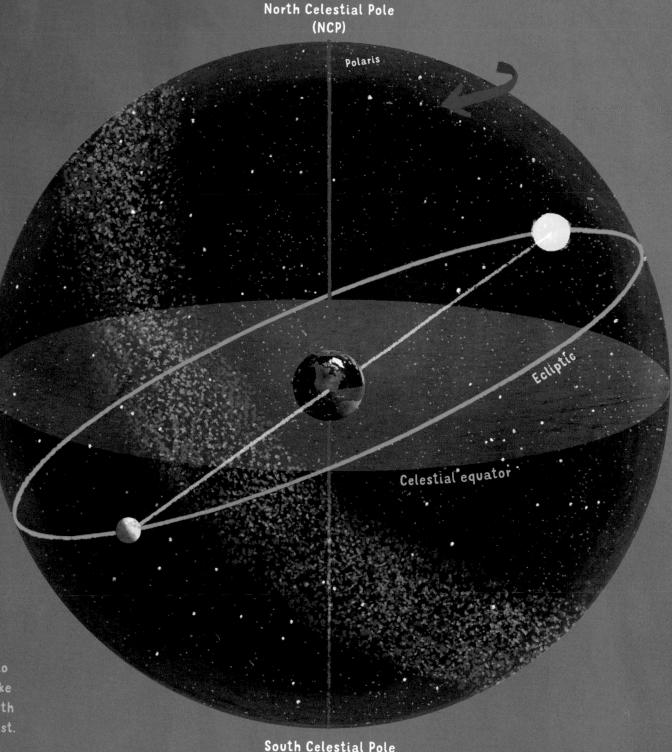

North Celestial Pole (NCP)

Polaris

Ecliptic

Celestial equator

South Celestial Pole (SCP)

All stars circle the pole, but stars closer to the pole make smaller circles. Some never cross the horizon. These are called **circumpolar** ("around the pole") stars.

Star Paths

A star's path in the sky depends on where it is on the **celestial sphere**. A star on the celestial equator rises due east, sets due west, and is above the horizon for 12 hours.

Aldebaran, the "Eye of the Bull" in the constellation Taurus, is north of the celestial equator. It rises in the northeast and sets in the northwest. In the Northern Hemisphere it is above the horizon for more than 12 hours.

Sirius, in the constellation of Canis Minor, is south of the celestial equator. It rises in the southeast and sets in the southwest; in the Northern Hemisphere it is above the horizon for less than 12 hours.

This may sound familiar! The Sun's path changes throughout the year; it is north of the celestial equator in June and south of it in December. It takes the same path as stars that are north and south of the celestial equator.

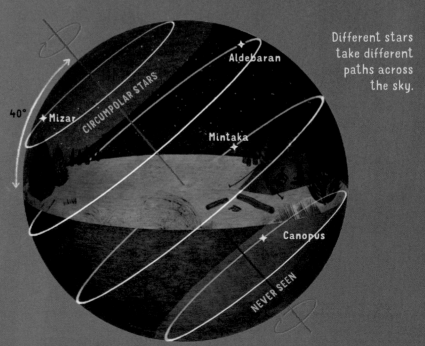

Different stars take different paths across the sky.

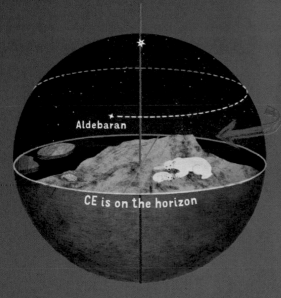

North Celestial Pole (NCP) is in zenith

Aldebaran

CE is on the horizon

THE SKY AT THE NORTH POLE

NCP is 40°
above horizon

Aldebaran

Sirius

CE is inclined at 50°

THE SKY AT 40° LATITUDE NORTH

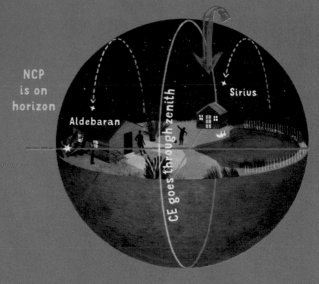

NCP
is on
horizon

Aldebaran

Sirius

CE goes through zenith

THE SKY AT THE EQUATOR

Where in the World? Changes in Latitude

We know that the Sun's path is different as seen from different places on Earth. This is true for the stars, too. Very far north, Polaris is high in the sky. Far south, it's low in the sky. Observers south of the equator can't see Polaris at all.

The angle between the North Celestial Pole and the horizon is equal to your (northern) latitude. In the Southern Hemisphere, the angle between the South Celestial Pole and the horizon is equal to your (southern) latitude.

The celestial equator is always 90 degrees from the poles. Even though you can't see it, the celestial equator is an arc stretching from east to west across the sky. It is tilted at an angle equal to 90 degrees minus your latitude. As the Earth spins, the stars spin around the poles, parallel to the celestial equator.

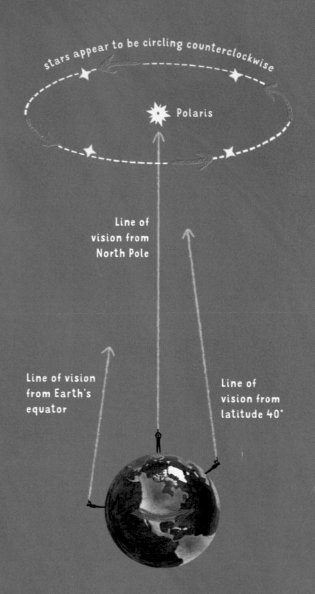

stars appear to be circling counterclockwise

Polaris

Line of vision from North Pole

Line of vision from Earth's equator

Line of vision from latitude 40°

New Place, New Stars

Some stars will never rise above your horizon. When ancient Greek astronomers traveled south to Egypt, they noticed that not only was Polaris lower in the sky, but they could see a new star that wasn't visible from Greece. They realized that the appearance of this new star, Canopus, meant that the Earth was a sphere.

What's UP, anyway? Your up is in a different direction from someone's up in a different place on Earth. In space there is no up or down because there is no gravity.

How Do Explorers Sail by the Stars?

When people first started sailing from Europe to North America, they used Polaris to help them find their way. They would sail west across the Atlantic Ocean, and when they were clear of any land they might crash into, they would measure the altitude of Polaris.

If they wanted to sail to Jamestown, Virginia, they would sail north or south until Polaris's altitude was equal to 37.2 degrees, the latitude of Jamestown. They kept Polaris on their right and sailed until they spotted land.

FIND NORTH & SOUTH USING THE STARS ✦

Like sailors throughout time, you can use the stars and constellations to help you find your way.

IN THE NORTHERN HEMISPHERE, Polaris is the North Star, always hanging in the sky above the North Pole. It is only medium-bright, but you can find it with the help of the Big Dipper in Ursa Major.

IN THE SOUTHERN HEMISPHERE, there is no "South Star" above the South Pole. Find south by picturing a line outward from the Southern Cross.

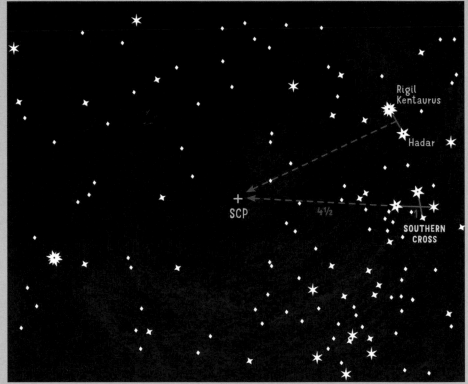

1. Find the Big Dipper. As one of the brightest asterisms (a group of stars that forms a picture but is not officially recognized as a constellation) in the sky, it is visible from most locations.

2. The two stars at the far end of the Big Dipper's bowl are called the Pointer Stars. Connect these two stars to draw a line that points directly to Polaris.

3. The distance from the Pointer Stars to Polaris is about 30 degrees (three fists) — five times the distance between the Pointer Stars.

How do you measure objects in the sky? With your cosmic protractors! See page 13.

1. The constellation known as the Southern Cross is circumpolar from most of the southern hemisphere. It's pretty low in the sky during October and November. The long bar of the cross points toward the South Celestial Pole (SCP).

2. Take the distance between the head (the star Gacrux) and foot (the star Acrux) of the cross, and extend it four and a half times its distance to reach the SCP. The SCP is about 25 degrees (two and a half fists) from the foot of the cross, or 30 degrees (three fists) from the head of the cross.

3. The two brightest stars in Centaurus, Rigil Kentaurus and Hadar, are called the Southern Pointers or just the Pointers. They point the way to the SCP. Trace an imaginary line between the Pointers, and then draw a line perpendicular to them. The SCP is about 30 degrees (three fists) from the Pointers.

WHY HUMANS INVENTED CONSTELLATIONS

People are good at seeing patterns. We look at clouds and see ships and sheep. We look at shadows and see scary monsters. We look at stars and see stories.

When early humans watched the night sky, they saw that stars rose and set just as the Sun, Moon, and planets did. They saw that stars stayed together in groups as they traveled across the sky.

Our ancestors connected the dots and made up stories about the pictures they saw in the stars — partly to help them remember what was where, and when. Many of the constellations that appear near each other are connected by a common story.

Orion, the Hunter, brandishes his weapon toward Taurus, the Bull.

ASTRONOMY NOTEBOOK

DIY CONSTELLATIONS ✦

Some people see Orion as a canoe, not a hunter. How might you see a star pattern in a new way?

1. Choose one of the seasonal sky maps from later in this chapter, and copy or trace only the stars into your notebook.

2. Look at it for a while. What patterns do you see in the stars? Connect the dots into shapes and name them.

3. Make up a story about your constellations.

THE ZODIAC AND THE ECLIPTIC

The ecliptic passes through a group of 12 constellations called the zodiac. They have names that may be familiar to you: Aries, Taurus, Gemini, Cancer, Leo, Virgo, Libra, Scorpius, Sagittarius, Capricornus, Aquarius, and Pisces.

The ecliptic also goes through Ophiuchus, the serpent bearer, but this constellation wasn't included in the zodiac because thirteen was considered to be an unlucky number.

The Sun spends about a month in each of the 12 zodiac constellations. The Moon and planets travel through the zodiac, too. They travel very slowly, though, so each night they appear at a fixed point. They rise and set just like stars in the same place on the celestial sphere (see page 91).

As the Earth revolves around the Sun, the Sun appears to move through the constellations of the zodiac. Here, the Sun is in the constellation Taurus.

Gemini · Taurus · Aries · Pisces · Aquarius · Capricorn · Cancer · Leo · Virgo · Libra · Scorpio · Sagittarius

SUN SIGNS

Astrology is a belief system based on the idea that the position of the stars and planets in the sky can predict what will happen in your life. The Babylonians, and later the Greeks, divided the ecliptic into 12 parts and named each part after the nearest constellation. These 12 divisions are called *signs*.

Your sun sign is the part of the ecliptic where the Sun was when you were born. A horoscope is a chart of where the Sun, Moon, planets, and constellations are at any particular moment.

STAR SEASONS

Chapter 3 told how our seasons change as we orbit the Sun. The stars have seasons, too. As the Earth travels around the Sun, the stars stay in the same place. So we see the stars at different times in different seasons.

In June, the Earth is on the same side of the Sun as Scorpius, but the opposite side from the constellation Taurus. For us, Taurus is up in the daytime sky, hidden by the bright light of the Sun. We can see Scorpius in the nighttime sky.

In December, the opposite is true: Scorpius is in the daytime sky, and we can see Taurus at night. As the Earth orbits the Sun, the stars rise four minutes earlier each day. Those four minutes add up to one hour earlier every two weeks and six hours earlier each season.

In Greek mythology, Scorpius and Orion were mortal enemies, so the gods had to separate them and placed them on opposite sides of the sky. (Taurus is next to Orion.)

WINTER SKY
Taurus the Bull, visible in December

The Pleiades, the Seven Sisters

Aldebaran, the Eye of the Bull

SUMMER SKY
Scorpius the Scorpion, visible in June

Antares, the Heart of the Scorpion

The Stinger [two stars]

MAKE YOUR OWN STAR WHEEL ★ ★

A star wheel (also called a *planisphere*) is a star map that you can adjust for different times of the year. Here's how to make your own to help you find stars and constellations.

YOU WILL NEED:

- Photocopier; or scanner and printer
- Scissors
- 2 sheets of card stock or a file folder
- Glue
- Paper fastener (optional)
- Markers or stickers

1. Look up your latitude on a map or a globe or online.

2. Photocopy or scan and print the star maps and holder, and cut them out. Cut out the holder along the curved line that is closest to your latitude.

3. Use one of the star maps to trace a circle onto the card stock, and cut it out. Glue a star map on to each side of the card-stock circle, making sure to line up the months on both maps.

4. Glue the holder to the card stock. Cut it out, making sure to follow the lines for your latitude. Don't forget to cut out the curved white areas — the windows!

5. Fold the holder across its "waist." Put dots of glue on the edges inside the dashed lines, fold the flaps over, and glue the front and back together.

6. If you live in the Northern Hemisphere, put the star wheel in the holder with the northern star map facing front. In the Southern Hemisphere, the southern star map should be in front.

7. Optional: Use the scissors to carefully poke a hole through the center of the circle and the center mark on the back of the holder. Line up the holes, and join them with the paper fastener.

8. Decorate the holder!

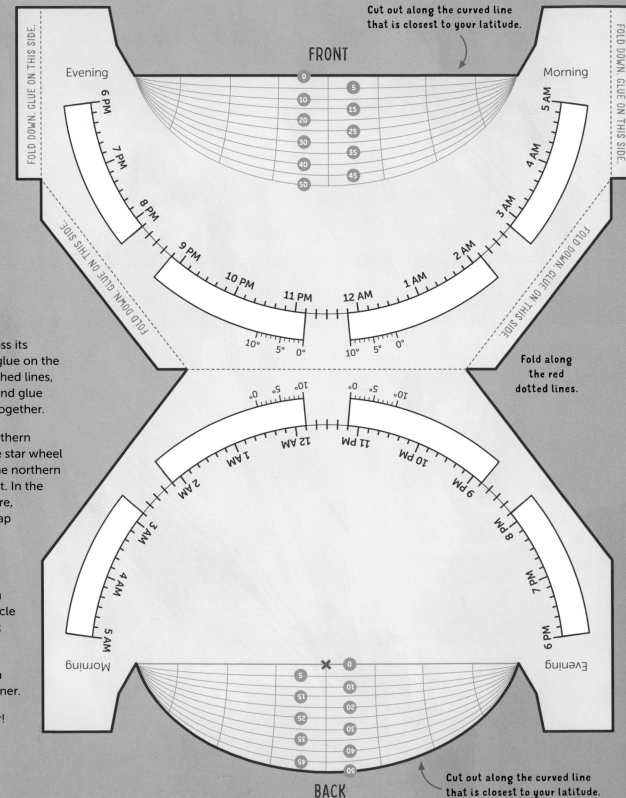

Cut out along the curved line that is closest to your latitude.

FOLD DOWN. GLUE ON THIS SIDE.

FRONT

Fold along the red dotted lines.

Cut out along the curved line that is closest to your latitude.

BACK

How Does It Work?

The curved edge of the holder represents Earth's horizon. In the Northern Hemisphere, the front is the northern horizon, and the back is the southern horizon. Write N in the center of the front horizon and S in the center of the back horizon.

If you are in the Southern Hemisphere, the front is the southern horizon, and the back is the northern horizon. Write S in the center of the front horizon and N in the center of the back horizon.

Looking at the northern sky, east is to your right and west is to your left. Looking south, east is to your left and west is to your right. Write E and W on the upper "ears" of the holder.

To see what the sky will look like at any date and time, line up the date on the star wheel with the time on the holder. While you're observing, make sure to double-check the time and date every once in a while. The wheel might rotate on its own!

The red circle is the ecliptic. Look for planets and the Moon here.

NORTHERN HEMISPHERE STAR MAP

SOUTHERN HEMISPHERE STAR MAP

Using Your Star Wheel

On a clear night, set the star wheel by lining up the date and the time in the windows. Take your red flashlight outside with you so you can read your star wheel in the dark. Make sure to wait at least five minutes for your eyes to adapt.

To find the constellations, face north and hold the north side of the star wheel — the side with N marked on it — in front of you. Larger symbols are brighter stars (see page 89). Can you match the constellations on the map with the stars in the sky?

Turn toward the east, and hold the star wheel so that E is on the bottom. Try matching the constellation with the eastern sky. Do the same for west (W) and south (S).

To find constellations overhead, look up at the zenith, and hold the star wheel over your head. Turn the holder so that it matches what you see in the sky.

Your star wheel will work for any time and any date, anywhere with a latitude within 5 degrees from your home.

SEASONAL SKY GAZING
Northern Hemisphere

The seasonal sky maps that follow show the constellations that rise in the east at sunset and are visible all night. You can see other constellations too, though. The constellations from the previous season are overhead and in the west at the beginning of the night. The constellations from the next season will rise after midnight.

You can explore this on your Star Wheel (see pages 98–99).

Northern Circumpolar Stars

The **BIG DIPPER** and **LITTLE DIPPER** are asterisms in the constellations of **URSA MAJOR** (the Big Bear) and **URSA MINOR** (the Little Bear). The Big Dipper points to the North Star, **POLARIS** (see page 90).

CASSIOPEIA (the Queen) is on the opposite side of Polaris from the Big Dipper. All through

NORTHERN CIRCUMPOLAR STARS

DRACO

URSA MINOR

CEPHEUS

Polaris

URSA MAJOR

CASSIOPEIA

PERSEUS

AURIGA

The Big Dipper turns around and changes position with the seasons.

DECEMBER

MARCH

JUNE

SEPTEMBER

the night, Cassiopeia and the Big Dipper spin slowly around Polaris.

Next to Cassiopeia is **CEPHEUS** (the King), a much fainter constellation. And wrapping around the Little Dipper and between all of these constellations is **DRACO** (the Dragon).

Pointer Stars

The two stars on the inner side of the Big Dipper's bowl make a line that points to **REGULUS**, the brightest star in **LEO**. Leo is one of the few constellations that actually looks like the thing it's supposed to be — a lion.

Use those same stars to point in the opposite direction to bright blue **VEGA**, one of the stars in the **SUMMER TRIANGLE** asterism (see page 108).

You can use the other stars in the Big Dipper to find **PEGASUS** (the Flying Horse), **AURIGA** (the Charioteer), and **GEMINI** (the Twins).

The handle of the Big Dipper makes an *arc* to **ARCTURUS**, the brightest star in the constellation **BOÖTES** (the Herdsman; see page 106). From Arcturus, *speed* on to **SPICA** (**speak**-uh or **spike**-uh), the brightest star in the constellation **VIRGO**.

Space Talk An **asterism** is a smaller group of stars, with its own name, within a larger constellation.

STARRY SIGNPOST

The Big Dipper is like a signpost that points the way to the North Star (Polaris), Leo, Gemini, and many other stars and constellations.

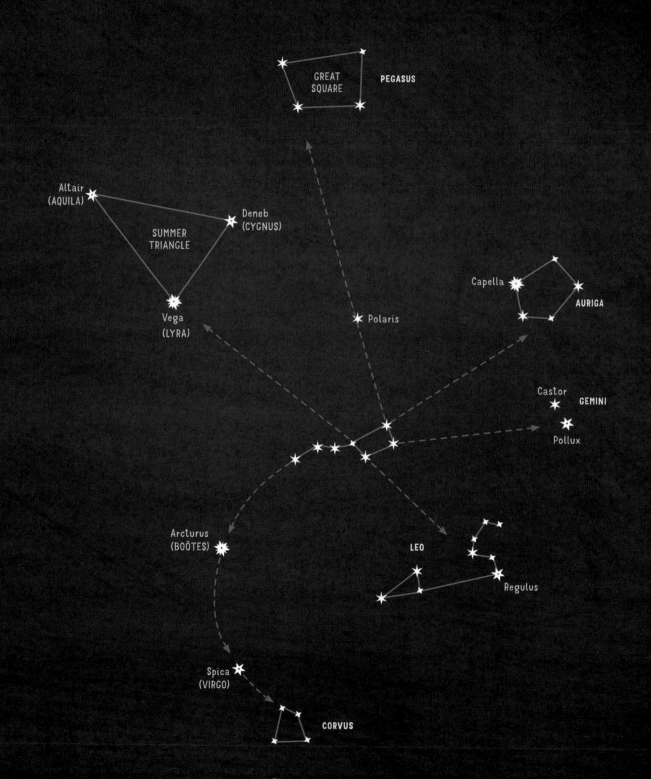

AUGUST, SEPTEMBER & OCTOBER STARS

The stars as seen at 9:00 p.m. on September 25, from a latitude of 40° north, looking east

CASSIOPEIA

M31

PEGASUS

GREAT SQUARE

ANDROMEDA

Algol

PERSEUS

Ecliptic

NORTH

EAST

SOUTH

The most noticeable constellations in September are Pegasus (the Flying Horse), Perseus (the Hero), and Andromeda (the Maiden).

The **GREAT SQUARE** asterism is the body of **PEGASUS**. **ANDROMEDA** is a line of three bright stars extending diagonally from one of the Great Square's corners. In Andromeda, you can find **M31**, the

Andromeda Galaxy, our Milky Way's nearest neighbor.

PERSEUS is between Andromeda and **CASSIOPEIA**. Some observers say the constellation resembles a shopping cart. The front of the shopping cart is the star **ALGOL**, the second brightest star in Perseus.

Algol has a faint companion star in orbit around it. Every

three days or so, Algol gets three times fainter for about 10 hours, when its light is blocked by the fainter star.

There is a beautiful pair of star clusters halfway between Perseus and Cassiopeia that's easy to see with binoculars. You can even see them with your naked eye if you are at a very dark observing site.

Clusters are groups of stars that were all born out of the same gas and dust cloud and are held together by gravity.

A GREEK STAR STORY

Cassiopeia and Cepheus were the king and queen of Ethiopia. Their daughter, Andromeda, was so beautiful that her mother couldn't help bragging about her. But when Cassiopeia claimed that Andromeda was more beautiful than the sea nymphs, there was trouble.

Poseidon, god of the sea, sent a monster to terrorize Ethiopia's coast. (Some say that Draco, the Dragon, was the sea monster; some say it was Cetus, the Whale.) The only way to save the country was to chain Andromeda to a rock and let the monster eat her.

Luckily for Andromeda, Perseus just happened to be flying by on his way home from killing Medusa, whose eyes could turn anyone to stone. He showed Medusa's head to the monster and turned it to stone. Then he unchained Andromeda from the rock and married her.

Pegasus, the Flying Horse, is in this group of constellations, too. Some legends say that Pegasus came to life when drops of Medusa's blood fell into the Mediterranean and mingled with the sea foam.

Algol (*r'as al-ghul*, or "head of the ogre," in Arabic) is said to be Medusa's head.

The stars as seen at 9:00 p.m. on December 25, from a latitude of 40° north, looking southeast

EAST

S. EAST

SOUTH

This is a great season for sky gazers! The bright, easy-to-find constellation of **ORION** (the Hunter) is visible the entire night. The two brightest stars in Orion are a nice contrast: **BETELGEUSE** is orange, and **RIGEL** is bluish white.

Orion's belt is the three equally spaced stars in the middle of the constellation. Hanging off the belt is Orion's sword, three much fainter stars. The middle "star" of the sword is actually **M42**, the **ORION NEBULA**, a great cloud of gas and dust where stars are forming. It's easily visible with binoculars.

In the Northern Hemisphere the bright stars in and around Orion are sometimes called the **WINTER HEXAGON**. Orion's belt points to **ALDEBARAN** in **TAURUS** (the Bull). The V of stars in Taurus is a star cluster called the **HYADES**.

About 10 degrees away is another star cluster, the **PLEIADES**, sometimes called the **SEVEN SIS-TERS**. There is a Greek myth about Orion chasing them. You can see this in the sky: as the stars rise and set, Orion follows the Pleiades from east to west. Many Aboriginal Australian groups tell a similar story.

In the opposite direction, Orion's belt points to **SIRIUS**, the eye of **CANIS MAJOR** (the Big Dog).

Nearby **PROCYON** is the brightest star in **CANIS MINOR** (the Little Dog).

A line drawn from Rigel to Betelgeuse points to the feet of **GEMINI** (the Twins). Gemini's brightest stars are **POLLUX** and **CASTOR**, which are the twins' names.

The final constellation in the hexagon is **AURIGA** ("or-**ee**-ga," the Charioteer), above Orion's head. It's a squashed pentagon containing the bright star **CAPELLA**.

A SOUTHERN AFRICAN STAR STORY

The Namaqua Khoikhoi people of southern Africa tell the story of a man who married the daughters of the sky god. One day, his wives sent him out to hunt for three zebras, but they gave him only one arrow. When he shot his arrow at the zebras, it missed. He was so embarrassed that he couldn't go home.

He still sits in the cold sky, his arrow lying useless on the ground. What's more, he can't retrieve it because a lion is crouched nearby.

The three zebras are Orion's belt, the fallen arrow is Orion's sword, and the hungry lion is Betelgeuse. The hunter and his wives are Aldebaran and the Pleiades in nearby Taurus.

FEBRUARY, MARCH & APRIL STARS

The stars as seen at 9:00 p.m. on April 15, from a latitude of 40° north, looking southeast

BIG DIPPER

LEO · Regulus

BOÖTES

Arcturus

CORONA BOREALIS · Gemma

HERCULES · M13

VIRGO

CORVUS

Spica

S. EAST

EAST

SOUTH

This time of year has a lot of constellations: first Leo (the Lion), then Boötes (the Herdsman), Corona Borealis (the Northern Crown), and Hercules.

LEO is a bright zodiac constellation. It's on the ecliptic, so the Moon and planets often pass through Leo on their journey around the Sun. **REGULUS**, Leo's brightest star, is bluish white when viewed through binoculars.

You already know how to find **ARCTURUS** by following the arc of the **BIG DIPPER'S** handle (see page 101). The constellation **BOÖTES**

The two little dots above the second *o* in Boötes mean that you pronounce both *o*'s: Bo-*o*-tays.

looks like an ice cream cone with Arcturus at its point.

Continue the arc from Arcturus to another bright star, **SPICA**. Blue Spica is the only bright star in the extremely faint zodiac constellation of **VIRGO** (the Virgin). Just to the right of Virgo, continuing in the arc, is a small diamond of medium-bright stars, **CORVUS** (the Crow).

To the left of the ice cream in Boötes is **CORONA BOREALIS** (the Northern Crown). It looks like a tiara, with the bright star **GEMMA** at its center.

To the left of Corona Borealis is **HERCULES**. Observers with 50 mm or stronger binoculars can find the star cluster **M13** in Hercules. M13 looks like a faint, fuzzy star, but it's really a **globular cluster,** a giant ball-shaped cluster of 400,000 stars orbiting the Milky Way.

A NATIVE AMERICAN STAR STORY

The Mi'kmaq (a Canadian First Nations people) tell the story of seven hunters and a bear. The hunters were all birds: Robin led the way, followed by Chickadee (with a cooking pot on her back) and Moosebird (also called gray jay). Then came Pigeon, Blue Jay, Saw-whet Owl, and Barred Owl.

They chased the bear all through the summer. Barred Owl got tired and went home. So did the others, one by one, until only Robin, Chickadee, and Moosebird were left.

This story reminds us of how the stars in the northern sky move. Robin, Chickadee, and Moosebird form the handle of the Big Dipper, and the bear is its bowl. The middle star of the handle,

Mizar, has a faint companion called Alcor – Chickadee's cooking pot!

During the night, the handle of the Big Dipper follows its bowl through the sky, just as the hunters follow the bear.

In summer, four hunters pass below the horizon, abandoning the hunt. In fall, the Dipper's bowl is closest to the horizon. Robin shoots the bear, turning the leaves on the trees below (and Robin's breast) red.

Corona Borealis (on the left) is the cave where Bear hibernates. He emerges in March – Spring in the Northern Hemisphere.

MAY, JUNE & JULY STARS

The stars as seen at 9:00 p.m. on July 25, from a latitude of 40° north, looking southeast

Vega
Epsilon Lyrae
LYRA
Deneb
CYGNUS   Albireo
SUMMER TRIANGLE
Altair   **AQUILA**
DELPHINUS
Antares
SCORPIUS
SAGITTARIUS

EAST

S. EAST

SOUTH

The three brightest stars in the sky from June to September are **ALTAIR** (in **AQUILA**, the Eagle), **DENEB** (the tail of **CYGNUS**, the Swan), and **VEGA** (in **LYRA**, the Harp). In the Northern Hemisphere these three stars together are called the **SUMMER TRIANGLE** (another asterism).

The head of Cygnus, the star **ALBIREO**, is near the center of the Summer Triangle. As seen through a telescope, Albireo is a beautiful orange and blue double star.

You can use Aquila to point at the faint constellation **DELPHINUS** (the Dolphin).

Lyra is an equilateral triangle attached to a parallelogram. The triangle is made up of Vega and two other stars. The one that's not attached to the parallelogram is **EPSILON LYRAE**, a double star easily seen in binoculars. What you will not see is that each "star" in the pair is actually a double star. Epsilon Lyrae's nickname is the Double Double.

SCORPIUS (the Scorpion) and **SAGITTARIUS** (the Archer) are two very bright zodiac constellations. They inhabit a bright part of the Milky Way, with plenty of stars, nebulae, and dark clouds to look at through binoculars (more about them on page 116).

The center of the Milky Way galaxy is in the direction of **SAGITTARIUS** (whose central stars are often called the **TEAPOT**).

The brightest star in Scorpius is **ANTARES**. Its name means "rival of Mars." People often confuse it with Mars because it's bright, red, and near the ecliptic.

A CHINESE STAR STORY

One of the oldest legends of China tells of Zhinu (織女; weaver girl), the seventh daughter of the goddess of heaven. She fell in love with Niulang (牛郎; cow herd), who was mortal. They married and had two sons. When the goddess found out about their marriage, she was angry. She sent Zhinu back to the sky to weave the clouds.

Zhinu and Niulang missed each other terribly, so Niulang rode to the sky on the hide of one of his cows, taking their sons with him. Before they could reunite, the goddess created the Silver River to keep them apart. Once a year, on the seventh day of the seventh month in the Chinese lunar-solar calendar, all the magpies in the world form a bridge across the river so that the family can be together. This day, known as Qixi (七夕, "chee-shee"), or the Double Seventh Festival, is sometimes called Chinese Valentine's Day.

Zhinu is the star Vega in the constellation Lyra. Parts of Lyra form her weaving loom. Niulang is the star Altair in Aquila. The two stars on either side of Altair are Zhinu and Niulang's sons. Aquila and Vega are on opposite sides of the Milky Way (the Silver River). The bright star Deneb in Cygnus is the middle of the magpie bridge.

SEASONAL SKY GAZING
Southern Hemisphere

SOUTHERN CIRCUMPOLAR STARS

The sky maps on pages 102–108 show the constellations as they look from the Northern Hemisphere. If you live in the Southern Hemisphere, some of those constellations will not be visible to you. Others will look upside-down! You can use your Star Wheel (see pages 98–99) instead of those maps.

Southern Circumpolar Stars

No star marks the **South Celestial Pole (SCP)**, and most of the constellations in this region are pretty faint. The exceptions, though, are extremely bright!

CRUX (the Southern Cross) is a small, bright constellation that points toward the SCP. If you are in a dark area, you will see that the Milky Way goes right through Crux. Crux is home to the **COALSACK NEBULA**, a dense dust cloud that looks like a dark patch in the Milky Way.

The constellation **CARINA** (the Keel) is the bottom piece of a larger ancient constellation, **ARGO NAVIS**. Its brightest star is **CANOPUS**, the second brightest star in the sky, after **SIRIUS**.

Carina is also home to the beautiful **CARINA NEBULA**, which is even larger and brighter than the

Orion Nebula. Check it out with binoculars; the **PINCUSHION** star cluster is in the same field of view.

The brightest stars of **CENTAURUS** (the Centaur) are circumpolar. **RIGEL KENTAURUS**, the foot of the centaur, is the brightest, and you can use it to

find the SCP (see page 94). **OMEGA CENTAURI** looks like a star, but it's actually the largest globular cluster in our galaxy.

The two most beautiful southern circumpolar objects are the **LARGE AND SMALL MAGELLANIC CLOUDS**. These are two small galaxies that orbit the Milky

Way and will eventually become part of our galaxy. They're called the Magellanic Clouds after Ferdinand Magellan (c. 1480–1521), a Portuguese explorer who saw them while making the first recorded voyage around the world.

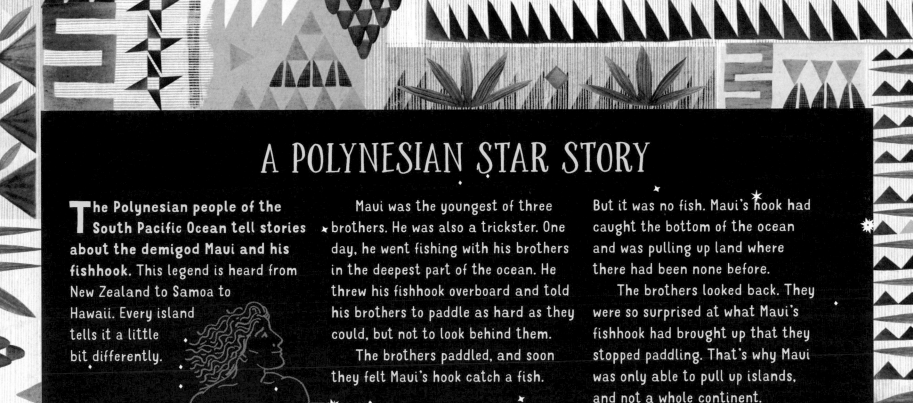

A POLYNESIAN STAR STORY

The Polynesian people of the South Pacific Ocean tell stories about the demigod Maui and his fishhook. This legend is heard from New Zealand to Samoa to Hawaii. Every island tells it a little bit differently.

Maui was the youngest of three brothers. He was also a trickster. One day, he went fishing with his brothers in the deepest part of the ocean. He threw his fishhook overboard and told his brothers to paddle as hard as they could, but not to look behind them.

The brothers paddled, and soon they felt Maui's hook catch a fish.

But it was no fish. Maui's hook had caught the bottom of the ocean and was pulling up land where there had been none before.

The brothers looked back. They were so surprised at what Maui's fishhook had brought up that they stopped paddling. That's why Maui was only able to pull up islands, and not a whole continent.

Maui's fishhook can be seen in the constellation Scorpius. In the sky, you can see that the fishhook has caught Pimoe (Sagittarius), a giant fish. The Summer Triangle (see page 108) is a coil of rope attached to the hook.

HOW STARS ARE BORN, LIVE, AND DIE

For most of their lives, stars are powered by nuclear fusion, which turns hydrogen into helium and releases heat and light (see page 52). When they are powered by hydrogen fusion, they are called main sequence stars. Our Sun is an example.

Main sequence stars are stable: the heat from fusion (pushing out) exactly balances gravity (pulling in).

A star like our Sun will live about 10 billion years. Less massive stars live longer; more massive stars burn hotter and live shorter lives.

Afterlife

When all the hydrogen in the star's core is turned to helium, the core starts to collapse and heat up. This pushes the star's atmosphere outward.

The core eventually becomes hot enough to fuse helium into carbon. The heat from this fusion can once again hold the star up against gravity. This helium-burning star is a **red giant**.

Supergiant or Dwarf?

When the red giant runs out of helium fuel its core will collapse again, and its outer layers will expand. At this point a very massive star takes a different course than a less massive one does.

A small- to medium-mass star like our Sun will continue to expand into a planetary nebula.

Trumpler 14, a very young star cluster, is still embedded in the cloud of gas and dust where it was born.

Its core collapses into a **white dwarf** — a star with the mass of the Sun, packed into the volume of the Earth. It has run out of all fuel and is at the end of its life.

When a star more than eight times the mass of the Sun runs out of fuel, its core will collapse and become hot enough to fuse heavier and heavier elements, all the way up to iron. This kind of star is called a supergiant.

The supergiant's iron core collapses very suddenly, causing the outer layers to explode in a supernova. A supernova explosion has so much energy that it can create elements heavier than iron.

Black Hole or Neutron Star?

After the explosion, if the star has less than three times the mass of our Sun, its core will become a neutron star, a super-dense star only about 15 miles (20 km) across. With more mass than that, the left-over core will become a **black hole**, a single point in space containing all the remaining mass of the star.

When a supernova explodes, the high-speed material it spews out can cause nearby gas clouds to collapse, starting the whole cycle of star formation over again!

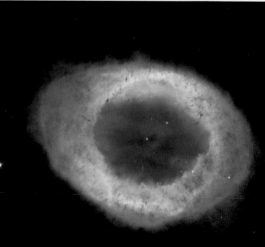

The Ring Nebula in the constellation Lyra is one of the best-known planetary nebulae. The "ring" is really a bubble of gas that used to be a star's atmosphere. The faint star at its center is a white dwarf.

We stars have made all of the elements in our universe that are heavier than helium.

LIFE STORY OF A STAR

A star is born in a gas cloud called a NEBULA.

WOOSH!

It is now called a PROTOSTAR. As gravity squeezes it, it heats up and starts to glow.

When its core gets hot enough, nuclear fusion begins, turning hydrogen to helium.

Now it's a MAIN SEQUENCE STAR, like our sun.

When all the hydrogen is used up, the core collapses and the atmosphere expands. Now it's called a RED GIANT.

The next phase depends how massive the star is.

Stars LESS THAN 8 times the mass of the Sun

A small- or medium-mass star will expand into a ring-shaped PLANETARY NEBULA of gases around a collapsing core.

The expanding nebula leaves behind a small, dense WHITE DWARF, which will eventually fade.

Stars MORE THAN 8 times the mass of the Sun

When the helium runs out, the core collapses and the atmosphere expands into a SUPERGIANT. Every time the core runs out of fuel, it collapses and fuses new, heavier elements, ending with iron.

BOOOM!

The iron core collapses and the outer layers explode in a SUPERNOVA.

A core LESS THAN 3 times the mass of the Sun

SWOOSH!

It will become a NEUTRON STAR, smaller and denser than a white dwarf.

A core MORE THAN 3 times the mass of the Sun

The entire mass of the core is compressed into a single point called a BLACK HOLE.

METEOR SHOWER

Shooting stars are not really stars; they're meteors — flashes of light caused by meteoroids entering Earth's atmosphere. The Earth sweeps up 100 tons of meteors every day, most of them too tiny to see. They hit the atmosphere at speeds of 27 to 56 miles per hour (43 to 90 km/hr — that's 12 to 25 m/sec!).

Most meteors are only a few millimeters across — about the size of a grain of sand — and are destroyed in our atmosphere before they reach the ground. Super-bright meteors come from meteoroids of around a centimeter across — the width of your pinkie.

During a meteor shower, most of the meteors seem to come from a single point in the sky, called the radiant. In reality, they are falling to Earth in parallel lines. For us on Earth it is similar to looking down a highway. The lines on the road are actually parallel, but they look like they're all coming from a single point in the distance.

You can see meteors any time. On a clear, dark night, away from city lights, you may see as many as five per hour. But several times a year, the Earth crosses over a comet's orbit, where there is a lot of rock and dust debris. Then there is a *meteor shower* – lots of meteors!

Meteor showers are named after the constellation where the radiant appears to be located.

Meteors are bright because, as they fall through our atmosphere, they collide with air molecules. These collisions heat up the meteors and cause the air molecules all along their path to glow. A larger meteor may heat up so much that it explodes and breaks into several pieces — then, it's called a bolide.

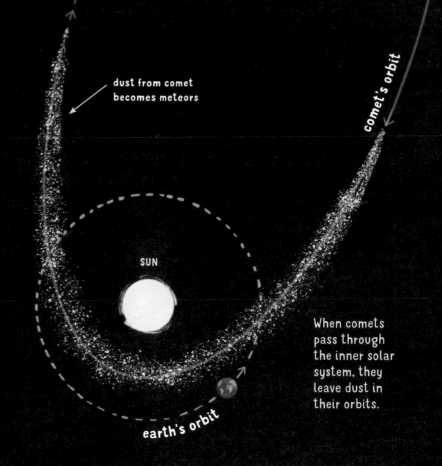

dust from comet becomes meteors

comet's orbit

SUN

When comets pass through the inner solar system, they leave dust in their orbits.

earth's orbit

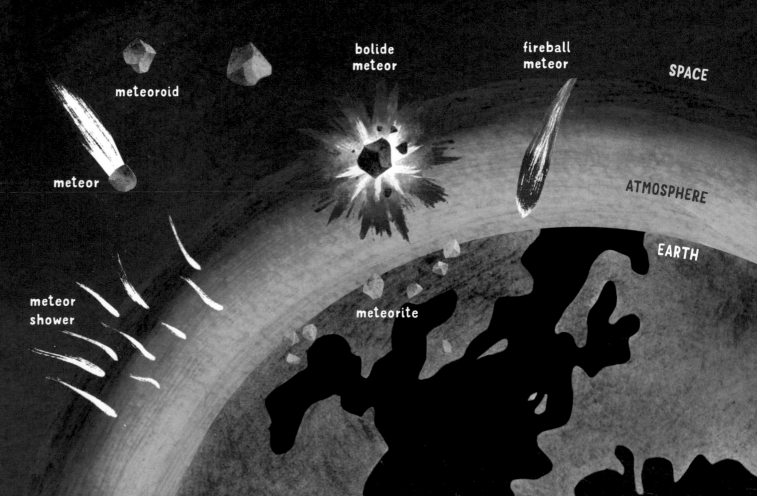

meteoroid

bolide meteor

fireball meteor

SPACE

meteor

ATMOSPHERE

meteor shower

meteorite

EARTH

A time-lapse photo of the Perseid meteor shower in August 2009. The meteors all seem to be coming from the same point in Perseus.

The two fuzzy "stars" near the center are the Double Cluster in Perseus (see next page for a "closeup").

The fuzzball at the top is the Andromeda Galaxy.

Radiant

PERSEUS

PLEIADES

Plane, Satellite, or Meteor?

We already know how to tell stars and planets apart. But there are other lights in the sky. Planes, satellites, and meteors all look like stars that move. Here's how to tell them apart.

★ Planes have blinking lights on them.

★ Satellites move fairly slowly, don't blink, and disappear before they cross the horizon.

★ Meteors race across the sky.

★ The International Space Station moves quickly and steadily.

The International Space Station is a high-flying laboratory run by the space agencies NASA (U.S.), Roscosmos (Russian), JAXA (Japanese), ESA (European), and CSA (Canadian). Find out when it will pass over you and where to look by visiting NASA's "Spot the Station" website.

How to Observe Meteor Showers

On page 119 is a list of the approximate dates of the brightest meteor showers throughout the year. Make sure to confirm each shower's exact date and time each year; you can find this information online. Also check the Moon phase: a bright gibbous or full Moon will make it hard to see fainter meteors. So will light pollution.

The best way to observe a meteor shower is with the naked eye. A telescope or binoculars will limit your field of view. Most meteors appear near the radiant, but they can come from any part of the sky.

Take a sleeping bag or lawn chair to lie down on. Dress for the weather, as you would for star gazing. Be patient! Twenty-five meteors per hour sounds like a lot, but that translates to one meteor every two minutes or so. And that number is on the higher side.

Every once in a while, a meteor shower will be stronger than usual. There were 10,000 meteors per hour during the 1833 Leonid meteor storm.

A CLOSER LOOK

Deep Sky Objects

Binoculars help you see stars and objects that are fainter than what you can see with your naked eye, because their lenses gather more light than your pupils do. You can also examine smaller objects more closely. Depending on the power of your binoculars, you may be able to observe open clusters, globular clusters, nebulae, and even galaxies.

OPEN CLUSTERS are groups of a few hundred stars that were formed from the same cloud of gas and dust and are kept together by gravity. The stars live together for a few hundred million years and then go their separate ways. They're scattered throughout the galaxy.

GLOBULAR CLUSTERS are also kept together by gravity, but they are much bigger — hundreds of thousands of stars — and stick together for life. They're called "globular" because they're shaped like globes. Globular clusters orbit the center of our Milky Way galaxy. They're much older than open clusters.

M 13, a globular cluster in the constellation Hercules

How to find the DOUBLE CLUSTER:

★ First find **CASSIOPEIA** in the northern sky. It looks like two mountains next to each other — one big and one small.

★ Find these two stars in Cassiopeia: **GAMMA CASSIOPEIAE**, the center star of Cassiopeia, and **RUCHBAH**, the peak of the smaller mountain.

★ A line drawn between these two stars points to **MIRFAK**, the brightest star in **PERSEUS**. Imagine a line between Gamma Cassiopeiae and Mirfak; the Double Cluster is halfway between them along this line.

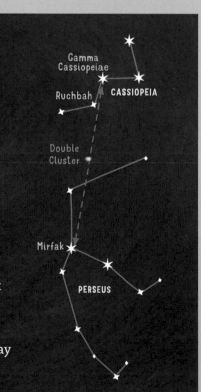

The Double Cluster

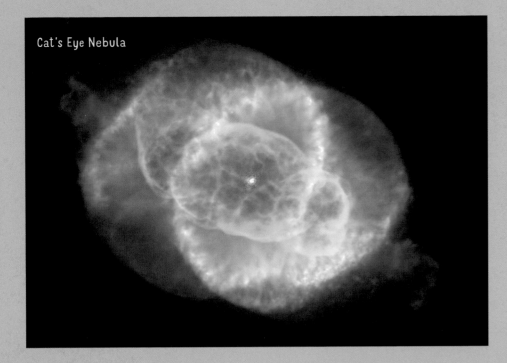

Cat's Eye Nebula

NEBULAE are clouds of gas and dust. The light from the stars in or near them makes the gas glow. Dark nebulae, like the Coalsack, are clumps of dust that block light from the stars behind them.

The three **GALAXIES** you can see with the naked eye (see page 20) are stunning through binoculars. You can see more galaxies through binoculars. They're listed in the Appendix, along with many other stars, clusters, and nebulae.

Stand back! The hottest stars are more than 70,000°F (40,000°C). The coolest are around 4,500°F (2,500°C).

Binocular Tips

Try pointing your binoculars anywhere in the Milky Way that looks cloudy; anything that looks like a cloud is usually a cluster of stars or a nebula. Scorpius, Sagittarius, and Cygnus are good places to look (as well as Centaurus, Carina, and Crux in the Southern Hemisphere).

One of the best ways to point at faint objects is a method called star hopping. Find an interesting object on a star chart, and look at what surrounds it in the sky. Point your binoculars at a star or constellation that you know, and hop to the object you don't know. For example, you can use the constellations of Cassiopeia and Perseus to find the Double Cluster, a pair of open star clusters.

STAR COLORS

If you look carefully, you may notice that many of the brighter stars in the sky are different colors. The color of a star tells you the temperature of its atmosphere: hot stars are blue, and cool stars are red. Red stars also tend to be fainter, unless they're red giants.

Rigel is BLUE

21,300°F (11,800°C)

Vega is WHITE

13,000°F (7,300°C)

10,000°F (5,500°C)

Betelgeuse is ORANGE

6,000°F (3,300°C)

Antares is RED

6,100°F (3,400°C)

FIND OUT MORE

Looking at the sky is fun, and we hope you do a lot of it. Here are a few more ways to learn about astronomy and get involved in science. Look on pages 124 and 125 for more information.

Science Museums

Just about every science museum has exhibits about astronomy. Many public libraries have passes that can get you into museums for free. You can check them out just like a book. If your family visits your local science museum more than once each year, consider becoming a member.

Libraries

Libraries have lots of books about astronomy and space science. Check the Resources section at the end of this book for other books to read. Your librarian can help you. Most libraries have special programs — see if your librarian could organize an astronomy program. Invite an astronomer to visit.

Planetariums

A planetarium is a room with a dome for its ceiling. A special projector projects images of the stars and constellations on the dome, so it looks like the night sky.

The projector can show what the sky looks like at different times of the year or from different places on Earth. It can show close-ups of your favorite objects in space.

The International Planetarium Society has an online tool to help you find a planetarium near you. There are also free planetarium apps you can install on a smart phone or computer to find out what's going on in your local sky. The WorldWide Telescope is an online planetarium — check it out!

The projector at Hamburg Planetarium in Germany can show the night sky on the dome above it.

Astronomy Clubs

Astronomy clubs are groups that meet every few weeks to learn about astronomy and telescopes. Many clubs welcome visitors to their meetings. Some even invite kids to join as junior members.

Most astronomy clubs have star parties, where members bring their own telescopes for people to look through. Or they invite local astronomy experts to talk about the cool science they do. Check out the list in the Resources section of this book to find the club nearest you.

Telescope Time

Some astronomy clubs and science museums have telescopes. Colleges, universities, and national observatories mostly use their telescopes to study the sky, but every once in a while they'll open their telescopes to the public. Check the website for the astronomy department of your nearest college to see when it offers public nights and other special events.

METEOR SHOWER CALENDAR

Here are the major meteor showers during the year, along with the best date and time to see them. The number of meteors per hour is for observers without light pollution. These dates can change by one or two days every year. Check with the American Meteor Society (www.amsmeteors.org) or other astronomy websites for precise timing as the event approaches.

NAME OF SHOWER	DATE	TIME	MAX NUMBER METEORS PER HOUR*	PARENT OBJECT	NOTES	CONSTELLATION
QUADRANTIDS (BOÖTIDS)	January 04	5:00 a.m.	120	2003 EH1 (asteroid)	Very short shower. Best seen from the Northern Hemisphere. Some fireballs.	Boötes
LYRIDS	April 22	4:00 a.m.	18	C/1861 G1 (Thatcher)	Lasts for three days. Best seen from the Northern Hemisphere. Some fireballs.	Lyra
ETA AQUARIIDS	May 07	4:00 a.m.	55	1P/Halley	Week-long shower. Best seen in the southern Tropics (equator to 25°S latitude).	Aquarius
SOUTHERN DELTA AQUARIIDS	July 30	3:00 a.m.	16	96P/Machholz 1 (parent object is not certain)	Two-week shower. Best seen in the southern Tropics. Fairly faint meteors.	Aquarius
ALPHA CAPRICORNIDS	July 27	1:00 a.m.	5	169P/NEAT	Best seen from the Tropics. Very weak shower with occasional fireballs.	Capricornus
PERSEIDS	August 12	4:00 a.m.	100	109P/Swift-Tuttle	Meteors are visible for about a week before and after August 12. Very strong shower.	Perseus
ORIONIDS	October 22	5:00 a.m.	25	1P/Halley	Usually a medium-strength shower, but can be stronger.	Orion
SOUTHERN TAURIDS	October 29	1:00 a.m.	5	2P/Encke	Weak but long-lasting (two months) showers that produce colorful fireballs. These two showers overlap with each other in late October and early November.	Taurus
NORTHERN TAURIDS	November 11	Midnight	5			
LEONIDS	November 18	5:00 a.m.	15	55P/Tempel-Tuttle	Has shown major outbursts in the past, but another is not due until 2099.	Leo
GEMINIDS	December 13	1:00 a.m.	120	3200 Phaethon (asteroid)	Best from the Northern Hemisphere. Strongest shower of the year. Bright, colorful meteors.	Gemini
URSIDS	December 22	5:00 a.m.	10	8P/Tuttle	Northern Hemisphere only. Unpredictable outbursts.	Ursa Minor

*If the maximum occurs when the radiant is at the zenith.

Source: American Meteor Society, https://www.amsmeteors.org/meteor-showers/2017-meteor-shower-list/ (2017 Meteor Shower List) and https://www.amsmeteors.org/meteor-showers/meteor-shower-calendar/ (Meteor Shower Calendar: 2019)

TRY IT

THROW A STAR PARTY ★ ★

A star party is a gathering of people to look at the sky. Star parties can be big or small, in the city or the country. They can include telescopes or naked-eye viewing. They can be organized by adults or kids (with a little adult help).

YOU WILL NEED:

- Red lighting (like the red flashlights on page 19)
- Constellation maps and/or a Star Wheel (see pages 98–99)
- Telescopes and binoculars (optional!)

For best viewing, your Star Party should happen in a dark place, such as a soccer field, pasture, park, or beach. Make sure everyone has a flashlight.

Every piece of equipment should have an experienced person in charge of it.

Have some kind of lighting so that people can be safe but still see the dark sky.

Make posters or flyers for your event, or invite people through social media. Flyers should clearly state the time and location, parking, and what will happen if it's cloudy. Include the contact information of an adult who is in charge of communication.

Some star parties happen during special events like eclipses or meteor showers, but you don't need a special occasion to share your love of the sky with your friends and neighbors.

Eclipses 2020–2030

DATE	TYPE		TOTAL OR ANNULAR	PARTIAL
2020 NOVEMBER 30	Lunar	Penumbral		Asia, Australia, Pacific, Americas
2020 DECEMBER 14	Solar	Total	Pacific, Chile, Argentina, Atlantic	Pacific, South America, Antarctica
2021 MAY 26	Lunar	Total	East Asia, Australia, Pacific, Americas	
2021 JUNE 10	Solar	Annular	Canada, Greenland, Russia	North America, Europe, Asia
2021 NOVEMBER 19	Lunar	Partial		Americas, Northern Europe, East Asia, Australia, Pacific
2021 DECEMBER 4	Solar	Total	Antarctica	Antarctica, Africa, Atlantic
2022 APRIL 30	Solar	Partial		Pacific, South America
2022 MAY 16	Lunar	Total	Americas, Europe, Africa	
2022 OCTOBER 25	Solar	Partial		Europe, Africa, Middle East, Asia
2022 NOVEMBER 8	Lunar	Total	Asia, Australia, Pacific, Americas	
2023 APRIL 20	Solar	Annular + Partial	Indonesia, Australia, Papua New Guinea	Asia, Indian Ocean, Oceania
2023 MAY 5	Lunar	Penumbral		Africa, Asia, Australia
2023 OCTOBER 14	Solar	Annular	United States, Central America, Colombia, Brazil	North America, South America
2023 OCTOBER 28	Lunar	Partial		Eastern Americas, Europe, Africa, Asia, Australia
2024 MARCH 25	Lunar	Penumbral		Americas
2024 APRIL 8	Solar	Total	Mexico, United States, Canada	North America
2024 SEPTEMBER 18	Lunar	Partial		Americas, Europe, Africa
2024 OCTOBER 2	Solar	Annular	Chile, Argentina	Pacific, South America
2025 MARCH 14	Lunar	Total	Pacific, Americas, Western Europe, West Africa	
2025 MARCH 29	Solar	Partial		Africa, Europe, Russia
2025 SEPTEMBER 7	Lunar	Total	Europe, Africa, Asia, Australia	
2025 SEPTEMBER 21	Solar	Partial		Pacific, Antarctica
2026 FEBRUARY 17	Solar	Annular	Antarctica	South America, Africa, Antarctica
2026 MARCH 3	Lunar	Total	East Asia, Australia, Pacific, Americas	

DATE	TYPE		TOTAL OR ANNULAR	PARTIAL
2026 AUGUST 12	Solar	Total	Arctic, Greenland, Iceland, Spain	North America, Africa, Europe
2026 AUGUST 28	Lunar	Partial		Eastern Pacific, Americas, Europe, Africa
2027 FEBRUARY 6	Solar	Annular	Chile, Argentina, Atlantic	South America, Antarctica, Africa
2027 FEBRUARY 20	Lunar	Penumbral		Americas, Europe, Africa, Asia
2027 JULY 18	Lunar	Penumbral		East Africa, Asia, Australia, Pacific
2027 AUGUST 2	Solar	Total	Morocco, Spain, Algeria, Libya, Egypt, Saudi Arabia, Yemen, Somalia	Africa, Europe, Asia
2027 AUGUST 17	Lunar	Penumbral		Pacific, Americas
2028 JANUARY 12	Lunar	Partial		Americas, Europe, Africa
2028 JANUARY 26	Solar	Annular	Ecuador, Peru, Brazil, Suriname, Spain, Portugal	North America, South America, Europe, Africa
2028 JULY 6	Lunar	Partial		Europe, Africa, Asia, Australia
2028 JULY 22	Solar	Total	Australia, New Zealand	Asia, Indian Ocean, Oceania
2028 DECEMBER 31	Lunar	Total	Europe, Africa, Asia, Australia, Pacific	
2029 JANUARY 14	Solar	Partial		North America
2029 JUNE 12	Solar	Partial		Arctic, Europe, Asia, North America
2029 JUNE 26	Lunar	Total	Americas, Europe, Africa, Middle East	
2029 JULY 11	Solar	Partial		South America
2029 DECEMBER 5	Solar	Partial		South America, Antarctica
2029 DECEMBER 20	Lunar	Total	Americas, Europe, Africa, Asia	
2030 JUNE 1	Solar	Annular	Algeria, Tunisia, Greece, Turkey, Russia, China, Japan	Europe, Africa, Asia, Arctic, North America
2030 JUNE 15	Lunar	Partial		Europe, Africa, Asia, Australia
2030 NOVEMBER 25	Solar	Total	Botswana, South Africa, Australia	Africa, Indian Ocean, Australia, Antarctica
2030 DECEMBER 9	Lunar	Penumbral		Americas, Europe, Africa, Asia

Source: Espenak, F. 2016, NASA Eclipse Web Site. https://eclipse.gsfc.nasa.gov/eclipse.html.
Permission is freely granted to reproduce this data when accompanied by an acknowledgment:
"Eclipse Predictions by Fred Espenak, NASA/GSFC Emeritus."

BINOCULAR OBJECTS

Here are some objects that are fun to look at through binoculars. We've listed the constellation each object is in, as well as the best date to view it and which hemisphere it is visible from. You can see the objects a month or two before and after their "Best Date." You will need to look online for charts that show you where most of these objects are.

MAGNITUDE (the object's brightness). Small magnitudes are bright objects; large magnitudes are dim. (See page 89 for more on magnitudes.)

BRIGHT SKY. This column tells you if you can observe the object *with* or *without* moonlight or light pollution.

DIFFICULTY. Start with objects that have a rating of one star, and go on to two and three stars when you are more experienced.

OBJECT	NICKNAME	TYPE	MAGNITUDE	CONSTELLATION	BEST DATE	HEMISPHERE	BRIGHT SKY	DIFFICULTY
MESSIER 41		open cluster	4.6	Canis Major	January 1	Southern	without	***
NGC 2451		open cluster	2.8	Puppis	January 14	Southern	with	*
MESSIER 46		open cluster	6	Puppis	January 14	Both	without	***
CALDWELL 96		open cluster	3.8	Carina	January 17	Southern	with	**
NGC 2547		open cluster	4.7	Vela	January 20	Southern	without	***
MESSIER 48		open cluster	5.5	Hydra	January 21	Both	without	***
MESSIER 44	Beehive Cluster	open cluster	3.7	Cancer	January 27	Both	with	**
CALDWELL 102	Theta Carinae Cluster, Southern Pleiades	open cluster	1.9	Carina	February 28	Southern	with	*
CALDWELL 92	Eta Carinae Nebula	nebula	1	Carina	February 28	Southern	with	*
δ (DELTA) CHAMAELEONIS		double star	4.5, 5.5	Chamaeleon	February 28	Southern	with	*
CALDWELL 91		open cluster	3	Carina	March 5	Southern	with	**
MELOTTE 111	Coma Berenices	open cluster	1.8	Coma Berenices	March 27	Northern	without	**
α (ALPHA) CRUCIS	Acrux	blue-white star with faint companion	1.4, 4.9	Crux	March 28	Southern	with	*
κ (KAPPA) DRACONIS		double star	3.9, 4.9	Draco	March 30	Northern	with	**
CALDWELL 94	Jewel Box	open cluster	4.2	Crux	April 4	Southern	with	*
CALDWELL 99	Coalsack Nebula	dark nebula		Crux	April 4	Southern	with	*
ζ (ZETA) URSAE MAJORIS	Mizar and Alcor	double star	2.2, 4.0	Ursa Major	April 12	Northern	with	*
CALDWELL 80	ω (omega) Centauri	globular cluster	3.6	Centaurus	April 13	Southern	with	***
α (ALPHA) VIRGINIS	Spica	blue-white star	0.97	Virgo	April 13	Both	with	*
MESSIER 3		globular cluster	6.2	Canes Venatici	April 17	Northern	without	**
β (BETA) CENTAURI	Hadar	blue-white star	0.6	Centaurus	April 23	Southern	with	*
α (ALPHA) BOÖTIS	Arcturus	yellow star	-0.05	Boötes	April 26	Both	with	*
τ (TAU) LUPI		double star	4.6, 5.0	Lupus	April 29	Southern	without	***
α (ALPHA) LIBRAE	Zuben elgenubi	double star	2.7, 5.2	Libra	May 5	Both	with	*
MESSIER 5		globular cluster	5.6	Serpens	May 13	Both	without	***
μ (MU) BOÖTIS	Alkalurops	double star	4.3, 6.5	Boötes	May 14	Northern	without	***
ω (OMEGA) SCORPII		double star	4.0, 4.3	Scorpius	May 25	Southern	with	**
δ (DELTA) APODIS		double star	4.7, 5.2	Apus	May 28	Southern	without	**
MESSIER 4		globular cluster	5.6	Scorpius	May 29	Southern	without	*
α (ALPHA) SCORPII	Antares	red star	0.91	Scorpius	May 30	Southern	with	*
MESSIER 13	Great Hercules Globular	globular cluster	5.8	Hercules	June 2	Northern	without	**
CALDWELL 76		open cluster	2.6	Scorpius	June 5	Southern	with	*

OBJECT	NICKNAME	TYPE	MAGNITUDE	CONSTELLATION	BEST DATE	HEMISPHERE	BRIGHT SKY	DIFFICULTY
μ (MU) SCORPII		double star	3.1, 3.6	Scorpius	June 5	Southern	with	*
ζ (ZETA) SCORPII		double star	3.6, 4.7	Scorpius	June 5	Southern	with	*
ν (NU) DRACONIS		double star	4.9, 4.9	Draco	June 14	Northern	without	**
MESSIER 6	Butterfly Cluster	open cluster	4.2	Scorpius	June 16	Southern	without	**
CALDWELL 86		globular cluster	5.6	Ara	June 17	Southern	without	***
IC 4665		open cluster	4.2	Ophiuchus	June 18	Both	without	***
MESSIER 7	Ptolemy's Cluster	open cluster	3.3	Scorpius	June 20	Southern	without	*
MESSIER 8	Lagoon Nebula	nebula	6	Sagittarius	June 22	Southern	without	***
MESSIER 24	Sagittarius Star Cloud	dense region of Milky Way	4.6	Sagittarius	June 25	Both	with	**
MESSIER 25		open cluster	6.5	Sagittarius	June 29	Both	without	***
MESSIER 22	Sagittarius Cluster	globular cluster	5.1	Sagittarius	June 30	Southern	without	***
α (ALPHA) LYRAE	Vega	blue star	0.03	Lyra	June 30	Northern	with	*
ε (EPSILON) LYRAE		double star	4.6, 4.7	Lyra	July 2	Northern	with	*
SCUTUM STAR CLOUD		dense region of Milky Way	–	Scutum	July 3	Both	with	*
δ (DELTA) LYRAE		double star	4.3, 5.6	Lyra	July 4	Northern	without	**
CALDWELL 93		globular cluster	5.4	Pavo	July 8	Southern	without	***
COLLINDER 399	Coathanger	asterism	3.6	Vulpecula	July 12	Northern	with	***
α (ALPHA) VULPECULAE		double star	4.6, 5.9	Vulpecula	July 13	Northern	without	***
ο (OMICRON) 1 CYGNI	30 & 31 Cygni	double star	3.9, 4.8	Cygnus	July 24	Northern	with	**
α (ALPHA) CAPRICORNI	Algedi	double star	3.7, 4.3	Capricornus	July 25	Both	with	**
β (BETA) CAPRICORNI	Dabih	double star	3.2, 6.1	Capricornus	July 26	Both	with	**
γ (GAMMA) EQUULEI		double star	4.7, 6.1	Equuleus	August 7	Both	without	***
μ (MU) CYGNI		double star	4.8, 6.9	Cygnus	August 16	Northern	with	***
π (PI) PEGASI		double star	4.3, 5.6	Pegasus	August 23	Northern	without	***
CALDWELL 106	47 Tucanae	globular cluster	4	Tucana	September 30	Southern	without	*
β (BETA) TUCANAE		double star	3.6, 5.1	Tucana	October 2	Southern	with	**
MESSIER 31	Andromeda Galaxy	galaxy	3.4	Andromeda	October 4	Northern	with	**
SMALL MAGELLANIC CLOUD	SMC	galaxy	2.7	Tucana	October 7	Southern	with	*
α (ALPHA) ERIDANI	Archernar	blue-white star	0.46	Eridanus	October 19	Southern	with	*
χ (CHI) CETI		double star	4.7, 6.8	Cetus	October 23	Both	without	**
CALDWELL 14	Double Cluster, χ (chi) Persei	open cluster	4.3	Perseus	October 31	Northern	with	**
CALDWELL 14	Double Cluster, h (eta) Persei	open cluster	4.3	Perseus	October 31	Northern	with	**
MELOTTE 20	Alpha α Persei Cluster	open cluster	1.2	Perseus	November 15	Northern	with	*
MESSIER 45	Pleiades	open cluster	1.6	Taurus	November 21	Northern	with	*
CALDWELL 41	Hyades	open cluster	1	Taurus	November 30	Both	with	*
κ (KAPPA) TAURI		double star	4.2, 5.3	Taurus	November 30	Northern	with	**
θ (THETA) TAURI		double star	3.4, 3.9	Taurus	December 1	Both	with	*
α (ALPHA) TAURI	Aldebaran	red-orange star	0.86	Taurus	December 2	Both	with	*
σ (SIGMA) TAURI		double star	4.7, 5.1	Taurus	December 3	Both	with	*
β (BETA) ORIONIS	Rigil	blue-white star	0.13	Orion	December 11	Both	with	*
LARGE MAGELLANIC CLOUD	LMC	galaxy	0.9	Dorado	December 13	Southern	with	*
MESSIER 38		open cluster	7.4	Auriga	December 15	Northern	without	***
MESSIER 36		open cluster	6.3	Auriga	December 16	Northern	without	***
MESSIER 42	Great Nebula in Orion	nebula	4	Orion	December 16	Both	with	*
42 & 45 ORIONIS		double star	4.6, 5.2	Orion	December 16	Both	with	*
CALDWELL 103	Tarantula Nebula	nebula	1	Dorado	December 17	Southern	with	*
MESSIER 37		open cluster	6.2	Auriga	December 20	Northern	without	***

BUYING BINOCULARS

Binoculars are usually described by their **magnification** and **aperture**. Magnification is a measure of how much bigger the binoculars make an object look. Aperture is the size of the objective lens (in millimeters).

A good binocular size for a beginning astronomer is 7 × 50 (pronounced "seven by fifty"). With them you can see planets, moon craters, and star clusters.

Binoculars are labeled with their magnification, aperture, and field of view.

Magnification

A pair of 7 × 50 binoculars has a magnification of 7 (it makes objects look seven times larger), and it has 50-millimeter (2-inch) objective lenses.

More magnification is not always better. With higher magnification, you see a smaller piece of the sky; and if your hands shake even a little bit, the image in your binoculars will shake a lot.

Aperture

The larger the aperture, the more light the binoculars let in and the better you can see faint objects. A larger aperture also helps you to see small details clearly.

When buying binoculars, get the largest aperture you can afford, up to 50 mm. Anything bigger than that will be too heavy to hold.

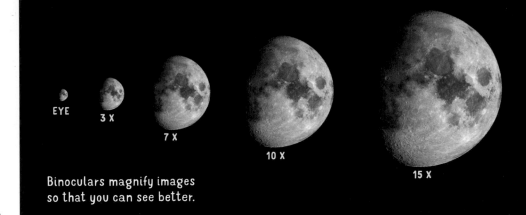

Binoculars magnify images so that you can see better.

Field of View

Binoculars are also labeled with their **field of view**. That's how much of the sky you can see through them, measured in degrees.

Other Tips

When shopping, look for:

* Binoculars with "coated optics" ("fully multi-coated" are best), specially treated lenses that reduce reflections and give you the clearest view
* Binoculars with a shock-proof rubber coating in case they are dropped

* A neck strap
* A case
* Lens caps

Clean your binocular lenses only if they are really dirty. To remove dirt, brush or blow dust off with a lens blower, bulb syringe, or camel-hair brush. Wipe off fingerprints or goo with a cotton swab dipped in isopropyl alcohol.

TRY IT
MAKE DEW SHIELDS ★★★

Prevent your binoculars from fogging up by adding dew shields.

1. Cut two rectangles from black craft foam. Each rectangle should be 6 to 10 inches (15 to 25 cm) long, and its width should be the circumference of your binoculars plus about 1 inch (2.5 cm).

2. Roll each rectangle into a long tube, with the extra inch of width as overlap. Glue down the overlap. (For fancy dew shields, attach self-sticking hook-and-loop tape to the overlap instead of gluing.)

3. Slide the tubes over the binocular barrels to protect the lenses from damp air.

RESOURCES

Books like This One

Croswell, Ken, *See the Stars*. Boyds Mills Press, 2000.

Rey, H.A., *The Stars*. Houghton-Mifflin, 1976.

Schneider, Howard, *Night Sky*. National Geographic, 2016.

Stott, Carole, *New Astronomer*. DK, 1999.

Books for Exploring Further

Carson, Mary Kay, *Exploring the Solar System*. Chicago Review Press, 2006.

Croswell, Ken, *Ten Worlds*. Boyds Mills Press, 2006.

Croswell, Ken, *The Lives of Stars*. Boyds Mills Press, 2009.

Dickinson, Terence, *Exploring the Sky by Day*. Camden House, 1988.

Dyson, Marianne, *Home on the Moon*. National Geographic, 2003.

Miller, Ron, *Stars and Galaxies*. Twenty-First Century Books, 2006.

Nichols, Michelle, *Astronomy Lab for Kids*. Quarry Books, 2016.

Read, John, *50 Things to See with a Small Telescope*. John Read, 2017.

Shore, Linda, Prosper, David & White, Vivian, *The Total Sky Watcher's Manual*. 2015, Weldon Owen, 2015.

Podcasts

Astronomy Cast
www.astronomycast.com
30 minutes of weekly astronomy current events

Planetary Radio
www.planetary.org/multimedia/planetary-radio
30–60 minutes of weekly planetary science and exploration news

StarTalk
www.startalkradio.net/
A two-minute update of what to look for in the sky tonight

Magazines

Astronomy

Sky & Telescope

Astronomy Current Events

Astronomy
www.astronomy.com/news

Astronomy Now
https://astronomynow.com/category/news

Astronomy Picture of the Day
https://apod.nasa.gov/apod/astropix.html

Science Daily
www.sciencedaily.com/news/space_time/astronomy

Sky & Telescope
www.skyandtelescope.com/astronomy-news

Space.com
www.space.com/news

Universe Today
www.universetoday.com

Websites with Times and Dates of Astronomical Events (including sunrise and sunset)

TimeandDate.com
www.timeanddate.com

In-The-Sky.org
https://in-the-sky.org

United States Naval Observatory
https://www.usno.navy.mil/USNO

EarthSky
https://earthsky.org

Aurora Information

Space Weather Prediction Center
National Oceanic and Atmospheric Administration
www.swpc.noaa.gov/products/aurora-30-minute-forecast
30-minute forecast, northern and southern lights

Spaceweather.com
https://spaceweather.com

Daily Image of the Sun

Big Bear Solar Observatory
www.bbso.njit.edu/cgi-bin/LatestImages

NASA Solar Data Analysis Center
https://umbra.nascom.nasa.gov/images

Spaceweather.com
https://spaceweather.com

Light Pollution

International Dark-Sky Association
www.darksky.org

INTERACTIVE MAPS OF LIGHT POLLUTION

The New World Atlas of Artificial Sky Brightness
Cooperative Institute for Research in Environmental Sciences at the University of Colorado Boulder
https://cires.colorado.edu/artificial-light

Blue Marble Navigator
https://blue-marble.de/nightlights/2012

Smartphone Apps for Measuring Sky Brightness

Dark Sky Meter
iOS only

Globe at Night

Loss of the Night

Where to Buy Astronomy Equipment Online

Anacortes Telescope & Wild Bird
www.buytelescopes.com

Astromart
www.astromart.com
Used equipment

B&H Photo and Video
www.bhphotovideo.com

Oceanside Photo & Telescope
https://optcorp.com

Orion Telescopes & Binoculars
www.telescope.com

Free Planetarium Apps

FOR COMPUTERS

Google Sky
www.google.com/sky

Stellarium
https://stellarium.org

Worldwide Telescope
www.worldwidetelescope.org

FOR SMARTPHONES

Night Sky
iOS only

SkEye
Really more for controlling telescopes than finding things on the sky. Android only

Sky Maps

Sky Safari

Sky View Free
iOS only

Star Chart
Android only

Starry
iOS only

Citizen Science

American Association of Variable Star Observers
www.aavso.org

American Meteor Society
www.amsmeteors.org

Association of Lunar and Planetary Observers
https://alpo-astronomy.org/

Astronomical League observing programs
https://cosmoquest.org

NASA Citizen Science
https://science.nasa.gov/citizenscience

SciStarter
https://scistarter.org

Zooniverse
www.zooniverse.org

Find an Astronomy Club Near You

ASP/NASA Night Sky Network
https://nightsky.jpl.nasa.gov/
USA only

Astronomical League
www.astroleague.org/
astronomy-clubs-usa-state
US only

Astronomy **Magazine**
www.astronomy.com/community/groups
International

Sky & Telescope **Magazine**
www.skyandtelescope.com/
astronomy-clubs-organizations
International

Online Astronomy Communities

Astronomers Without Borders
www.astronomerswithoutborders.org

Astronomical League
www.astroleague.org

Astronomical Society of the Pacific
www.astrosociety.org

SkyMaps.com
http://skymaps.com/downloads.html
Free monthly sky maps

Finding Satellites

NASA SkyWatch
https://spaceflight.nasa.gov

NASA Spot the Station
https://spotthestation.nasa.gov/home.cfm
You can even get automatic text notifications when the ISS is going to be overhead

Visual Satellite Observer
www.satobs.org

APPS

Heavens Above
www.heavens-above.com
iOS and Android

ISS Detector Satellite
www.issdetector.com
Android only

Cool Sites to Explore

Lunar Reconnaissance Orbiter Camera
www.lroc.asu.edu/images
Images of the Moon from the Lunar Reconnaissance Orbiter

MeteorShowers.org
www.meteorshowers.org
Positions of the major meteor showers

My NASA Data
https://mynasadata.larc.nasa.gov

NASA's Eyes
https://eyes.nasa.gov
Current locations of solar system spacecraft

Polynesian Voyaging Society
www.hokulea.com

Sky & Telescope
www.skyandtelescope.
com/astronomy-resources/
how-to-make-a-sundial
A simple paper sundial

Eclipse Information

American Astronomical Society
https://eclipse.aas.org/resources/
solar-filters
List of safe eclipse glasses

MrEclipse.com
https://mreclipse.com

NASA Eclipse Website
https://eclipse.gsfc.nasa.gov/eclipse.html
Solar and lunar eclipses

Weather Forecasts All over the World

Bureau of Meteorology
www.bom.gov.au
Australia

Clear Sky Chart
www.cleardarksky.com/csk
Astronomical forecast for Canada, the United States, and parts of Mexico

Environment Canada Weather
https://weather.gc.ca

India Meteorological Department
https://mausam.imd.gov.in

MET éireann
www.met.ie
Ireland

Met Office
www.metoffice.gov.uk
United Kingdom

Met Service
www.metservice.com
New Zealand

National Weather Service
www.weather.gov
United States

South African Weather Service
www.weathersa.co.za

GLOSSARY

altitude. An object's angle of elevation above the horizon

angle. The space between two lines that meet at a point, measured in degrees

annular eclipse. A type of solar eclipse where we can still see a thin ring, or annulus, of the Sun around the edge of the Moon

aperture. The opening that allows light to enter a telescope or binoculars

asterism. A group of stars that form a picture or pattern but are not officially recognized as a constellation

asteroid. A rocky body in the solar system found in the asteroid belt

asteroid belt. The region in the solar system between the orbits of Mars and Jupiter

atom. The tiny particles that make up elements

AU (astronomical unit). The average distance from Earth to the Sun — about 93 million miles (150 million km)

aurora. The northern (or southern) lights, caused by the solar wind interacting with Earth's atmosphere

axis of rotation. An imaginary line going through the center of an object, around which it rotates

azimuth. The angle of an object around the horizon from one of the cardinal directions (north, south, east, and west)

black hole. A large mass compressed to a single point. Its gravity is so strong that light can't escape it.

celestial sphere. The imaginary sphere of the sky surrounding Earth

circumpolar. Describes stars that don't rise or set but appear to circle around the north or south celestial pole

coma. A comet's atmosphere

comet. A large chunk of ice and rock in the solar system whose long, skinny orbit brings it close to the Sun

compound. A substance formed by two or more elements joined together

conjunction. When two objects in the solar system appear a few degrees apart in the sky

constellation. A group of stars that form a picture or pattern. There are 88 official constellations.

core. The central region of a spherical object like a star or planet

crust. The outermost layer of a moon or planet

differentiated. Describes planets and moons with more dense material at their core and less dense material on their surface

dwarf planet. An object orbiting a star. A dwarf planet is massive enough to be spherical but has not cleared its orbit of planetesimals.

ecliptic. The imaginary line in the sky that marks the plane on which all the planets orbit the Sun. As seen from Earth, the Sun appears to follow the ecliptic across the sky throughout the year.

electron. A subatomic particle with a negative charge

element. A pure substance containing only one kind of atom. The elements are the building blocks that make up all other matter in the universe.

equator. The imaginary circle on a sphere halfway between its poles

equinox. One of two days when the Sun is directly above the Earth's equator. They are March 20 and September 22 (give or take a day).

exoplanet. A planet that orbits a star other than our Sun

field of view. How much of the sky you can see, measured in degrees

fusion. The union of the nuclei in atoms to form heavier nuclei, which can result in the release of lots of energy. In the core of a star, hydrogen nuclei are fused to form helium, and the resulting energy powers the star.

galaxy. A huge system made of gas, dust, and billions of stars, held together by gravity

globular cluster. A ball-shaped cluster of hundreds of thousands of stars that orbits a galaxy

gravity. A force that pulls objects toward each other. Any object with mass has a gravitational pull. The more mass an object has, the greater its gravity.

greatest elongation. The point where Mercury or Venus is farthest from the Sun, as seen in the sky from Earth

horizon. An imaginary circle surrounding you at eye level

inferior planet. A planet that is closer to the Sun than the Earth is

ion. An atom with a positive or negative electric charge because it has lost or gained electrons

jovian. Like Jupiter

Kuiper belt. A doughnut-shaped region of icy planetesimals 30 to 50 AU from the Sun. Named for Gerald Kuiper ("coy-per")

latitude. Angular distance north or south of the Earth's equator

light pollution. Artificial light at night that makes the sky bright

light-year. The distance that light travels in one year, about 5.88 trillion miles (9.46 trillion km)

lunar eclipse. When the Earth's shadow blocks out the full Moon, as seen from Earth

magnification. How much bigger an object looks

main sequence star. A star that is fueled by nuclear fusion

mantle. The semi-liquid region inside a planet between the core and the crust

mare. A smooth, dark plain on the Moon's surface. Pronounced "**mar**-ay." Plural: maria

mass. A measure of the amount of matter an object contains

meridian. An imaginary line in the sky connecting the north point on the horizon, the zenith, and the south point on the horizon

meteor. A flash of light caused by a meteoroid when it enters the atmosphere of a planet or moon

meteorite. A meteor that lands on the surface of a planet or moon

meteoroid. A small rocky body moving through outer space

Milky Way. Our home galaxy. We see it as a misty stripe through the night sky.

nadir. The imaginary point at the bottom of the sky, opposite the zenith — right beneath your feet

nebula. A cloud of gas or dust in space. Plural: nebulae

neutron. A particle with no charge found in an atom's nucleus

North Celestial Pole (NCP). An imaginary point in the sky directly above Earth's North Pole. As the Earth rotates, the NCP stays still and all the stars appear to circle around it.

nuclear fusion. *See* fusion

nucleus. (1) The center of an atom, containing protons and neutrons. (2) The central, solid part of a comet. Plural: nuclei ("**new**-klee-eye")

objective. In a telescope or binoculars, the lens or mirror that focuses the light of a distant object

Oort cloud. Sphere-shaped region of icy planetesimals that extends from 2,000 to 50,000 AU from the Sun. Named for Jan Oort ("ort")

opposition. When an object is on the opposite side of the Earth from the Sun

orbit. The path that one object takes around another, like the Moon's orbit around the Earth. *Orbit* is also a verb: the Moon orbits the Sun.

partial eclipse. When only part of the Sun or Moon is blocked during an eclipse

penumbra. The outer edge of a shadow, which is dim and fuzzy

phase. One of the stages in a repeating pattern

plane. A flat surface

planet. An object orbiting a star. A planet is massive enough to be spherical and does not share its orbit with any other objects.

planetesimal. A chunk of rock a few miles (km) across. Many planetesimals can come together to form a planet

proton. A particle with a positive charge, found in an atom's nucleus

protoplanet. A large mass of matter orbiting the Sun or a star, and forming into a planet

protostar. A mass of gas that has not yet started fusion in its core

red giant. A star that has run out of hydrogen in its core and is now fusing helium into carbon

solar eclipse. When the new Moon blocks out the Sun, as seen from Earth

solar wind. The steady stream of protons and electrons that extends from the Sun's corona and throughout the solar system

solstice. The day when the Sun is at its northernmost (June 21) or southernmost point (December 21)

South Celestial Pole (SCP). An imaginary point in the sky directly above Earth's South Pole. As the Earth rotates, the SCP stays still and all the stars appear to circle around it

sunspot. A relatively cool region on the Sun's surface

superior planet. A planet that is farther away from the Sun than the Earth is

terrestrial. Like Earth

total eclipse. When the entire Sun or Moon is blocked during an eclipse

totality. The phase of an eclipse where the Sun or full Moon is completely covered

transit. (1) When an object in the sky crosses the meridian. (2) When a planet passes between the Earth and the Sun, partially blocking the Sun as seen from Earth

Trans-Neptunian Object (TNO). An object in our solar system beyond the orbit of Neptune

tropics. The region on Earth between 23.5 degrees north latitude and 23.5 degrees south latitude

twilight. The time right after sunset or before sunrise when the sky is not bright but softly glows

umbra. The dark center part of a shadow, where all light is excluded

vaporized. Transformed from a liquid to a gas

white dwarf. A star that has run out of all fuel. The last stage of a small- to medium-size star's life

zenith. The imaginary point in the sky directly above your head

zodiac. The 12 constellations that the Sun and planets travel through during a year. In order from the March Equinox, they are Pisces Aries, Taurus, Gemini, Cancer, Leo, Virgo, Libra, Scorpius, Sagittarius, Capricornus, and Aquarius.

INDEX

Acknowledgments

A book like this isn't made in the vacuum of space; there are a lot of people to thank.

First, of course, my incredibly patient family. My supportive colleagues at Smith College: James Lowenthal, Suzan Edwards, and Julie Skinner Manegold. Joel Weisberg, my first astronomy teacher and original source of the term "cosmic protractor."

My editor at Storey, Deb Burns, without whom this book literally would not exist, and book designer Jess Armstrong, who makes me look good. My critique groups, who made my words so much better: Carol, Kate, Lisa, Marie, Marjorie, MJ, Sheila, and Stephanie. Eric Jensen, who checked all the astronomy facts. (Any remaining errors are mine.)

Additional thanks go out to all the sky gazers I've met throughout my career: those who taught me, those I taught, and those who watched the skies with me.

Found all the planets?

Here's where they are:
Mercury, page 7; Venus, page 8; Earth, page 9; Mars, page 11; Jupiter, page 25; Saturn, page 43; Uranus, page 82; Neptune, 125.

EXPLORE THE NATURAL WORLD
WITH MORE BOOKS FROM STOREY

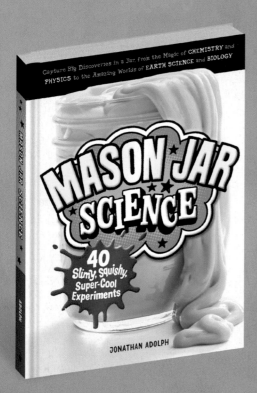

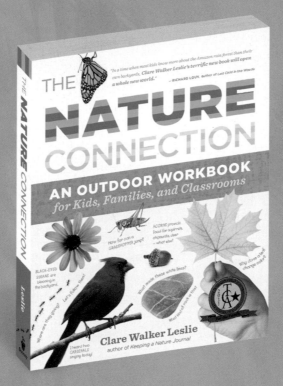

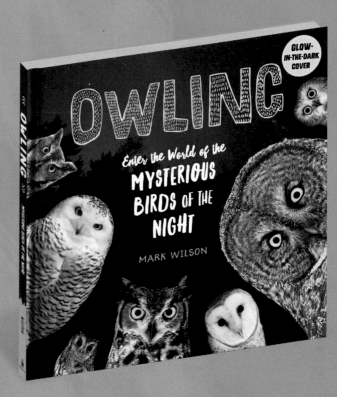

by Jonathan Adolph

From water fireworks and soda stalactites to a caterpillar hatchery, a balloon barometer, and much more, you can conduct these 40 fun, foolproof, and fascinating science experiments in a glass canning jar.

by Clare Walker Leslie

Dozens of interactive projects make every season a new discovery, with fun prompts to record daily sunrise and sunset times, draw a local map, keep a moon journal, or collect leaves to identify.

by Mark Wilson

Dramatic photos reveal the lives of 19 North American owl species as they nest, fly, and hunt. You'll learn about these birds' silent flight, remarkable eyes and ears, haunting calls, and fascinating behaviors.

Join the conversation. Share your experience with this book, learn more about Storey Publishing's authors, and read original essays and book excerpts at storey.com.
Look for our books wherever quality books are sold or call 800-441-5700.